REINHOLD MESSNER
My Life at the Limit

REINHOLD MESSNER
My Life at the Limit

Interviewed by Thomas Hüetlin
Translated by Tim Carruthers

A LEGENDS AND LORE TITLE

MOUNTAINEERS
BOOKS

Mountaineers Books is the nonprofit publishing division of
The Mountaineers, an organization founded in 1906 and
dedicated to the exploration, preservation, and enjoyment
of outdoor and wilderness areas.

**MOUNTAINEERS
BOOKS**

1001 SW Klickitat Way, Suite 201 • Seattle, WA 98134
800.553.4453 • www.mountaineersbooks.org

Mein Leben am Limit: Eine Autobiographie in Gesprächen mit Thomas Hüetlin © 2004,
2014 by Piper Verlag GmbH, München, published under the imprint Malik
English-language translation © 2014 by Mountaineers Books

Printed in the United States of America
17 16 15 14   1 2 3 4 5

Copy Editor: Elizabeth Johnson
Design and Layout: Emily Ford
Photos: Reinhold Messner Archives

Cover Photograph: www.arneschultz.com
Frontispiece: © David C. Tomlinson/Getty Images. *Dolomites in Parco Naturale Puez-Odle*

German text credits: Horst Höfler and Reinhold Messner: *Hermann Buhl: Am Rande des
Möglichen.* AS Verlag, Zürich, 2003; Alexander Langer: private writings; Ralf-Peter Märtin:
Nanga Parbat: Wahrheit und Wahn des Alpinismus. Berlin Verlag, Berlin, 2002; Christoph
Ransmayr: *Die Schrecken des Eises und der Finsternis,* S. Fischer Verlag, Frankfurt a. M.,
1996; Raoul Schrott: *Gilgamesh.* Hanser Verlag, Munich, 2002.

Library of Congress Cataloging-in-Publication Data

Messner, Reinhold, 1944-
  [Mein Leben am Limit. English]
  Reinhold Messner : my life at the limit / interviewed by Thomas Huetlin ; translated by Tim
Carruthers.
      pages cm
  "A Legends and Lore Title."
  ISBN 978-1-59485-947-2 (cloth)— ISBN 978-1-59485-852-9 (pbk.) — ISBN 978-1-59485-
853-6 (ebook)
  1. Messner, Reinhold, 1944- 2. Messner, Reinhold, 1944—Interviews. 3. Mountaineers—
Italy—Biography. 4. Travelers—Italy—Biography. 5. Extreme environments. I. Hüetlin,
Thomas, 1961- interviewer. II. Title.
  GV199.92.M47A313 2014
  796.522092—dc23
  [B]
                              2014018059

 Printed on recycled paper

ISBN (hardback): 978-1-59485-947-2
ISBN (paperback): 978-1-59485-852-9
ISBN (ebook): 978-1-59485-853-6

CONTENTS

PUBLISHER'S NOTE

In 2004 and then again in 2014, legendary climber Reinhold Messner sat down with journalist Thomas Hüetlin for a series of wide-ranging conversations. In the resulting exchange, Messner reflects on how his life experiences led him from his childhood in postwar Italy to becoming the world's most renowned mountaineer to his current passion: curating a group of six museums about mountains, collectively known as the Messner Mountain Museums.

This new book, *Reinhold Messner: My Life at the Limit*, assembles and distills those conversations in an interview format. As you read, "M" indicates Reinhold Messner is speaking, while "H" indicates Thomas Hüetlin.

CHAPTER I
DEFYING GRAVITY

1949–1969

*Thoughts worth thinking should not just
be understood; they should be experienced.*

—Count Harry Kessler,
May 1896

A CHILDHOOD ON THE ROCKS

I have been a rock climber for as long as I can remember.

As a boy, I didn't just climb on the cliffs of the local Geisler peaks; I climbed on the house-sized boulders at the edge of the forest, on the walls of ruined buildings, and on the cemetery wall during school break time. Most of all, I dreamed of climbing.

Thinking I was a little more able than I actually was, I imagined myself climbing steeper and steeper rock faces—until it seemed that no route was impossible for me. In my mind's eye I managed to make a series of first ascents on the highest faces in the Dolomites, and on the Eiger, Kilimanjaro, and Aconcagua.

I was also attending school and, like all my brothers and my sister, I had to help look after the chickens at home; it was these that made it possible for my parents to feed nine children. My father was a village schoolteacher. He was also my first climbing mentor, but by the age of ten or twelve I was climbing harder routes with my younger brother, and we were soon to enter a realm that belonged to us alone.

During my last few years of school, I came to realize that my path to knowledge would not lead me to libraries, professors, universities, and studies. My path to knowledge was through living life and experiencing reality. I could learn plenty secondhand, but nothing was ever to surpass the experiences I had in the wilderness. All my knowledge of social, scientific, and religious issues has been acquired through personal experience.

This is one of the reasons why in later life I kept forcing myself to organize the next expedition, the next big trip. How often on an expedition have I told myself, "That's enough!" and then a few weeks later when the effort, worry, and hardship were forgotten, I began dreaming about a new journey, planning a new climb. Pretty soon I'd be off again. And once again, it would be dangerous.

I never intended to risk my neck, but I knew that if I were ever to stop dreaming or traveling I would be old. And that would drive me to despair.

11

It was midday, and the four of us were sitting on a sharp, rocky ridge on the Secéda in the Geisler Range—my father, two of my brothers, and me. Above us was the Kleine Fermeda. The south face was bright in the sun. It looked steep but well featured; the line of the route was logical. A few scattered clouds hung like cotton wool over the peaks of the southeastern Dolomites, which rose above the Puez Plateau. That meant the weather was going to stay fine.

It wasn't curiosity or high spirits alone that made me keep looking up at the face above us; it was more than that. Perhaps it could be described as the desire to measure up.

Since my father had no objections, I set off alone without a rope. I scrambled down a rocky terrace for a bit, then climbed up and rightward. The rock was quite smooth, and it wasn't particularly steep at the start, although the cliff dropped away abruptly beneath me. I didn't look down, only at the rock face in front of me, taking it step-by-step, hold-by-hold. This was exactly what I wanted to do, to climb without distractions, following my instincts, finding the route as I went. I felt proud of myself.

I had now arrived at the crux and took a good look at the vertical wall above me. After spying out a series of handholds and footholds, I started climbing again.

Everything else was forgotten; it was just handholds and footholds and unrestrained movement. I was oblivious to everything around me. I might have hesitated briefly, looked down at my feet and seen the abyss that dropped 300 meters to the green alpine pastures below. After a few meters the climbing got easier again, and I was soon standing, carefully, on the south summit before scrambling up loose rock to the main summit. Looking north, I could see right down to the pastures of the Gschmagenhart Alm, where we had set out from that morning. To the south I could see all the famous Dolomite peaks, from the Langkofel to Sass Songher, with the Marmolada, Monte Pelmo, and the Civetta behind.

Climbing for me was more than a sport. Danger and difficulty were part of the game, together with adventure and exposure. Climbing a big route means total commitment. It requires total reliance on yourself for maybe several days as you try to unlock its secrets.

Climbing is all about freedom, the freedom to go beyond all the rules and take a chance, to experience something new, to gain insight into human nature. And there is always more than one answer to a question, more than one story behind every experience. For me, imagination is more important in climbing than muscle or daredevil antics. It is worth more than technology. The development of the person is more important than having bolt ladders everywhere. There are few treasures to be found in bolt-protected climbs. We need to protect the diversity of climbing, not every meter of rock.

H: You grew up in Villnöss, a valley in South Tyrol at the foot of the Geisler Range that has remained pretty much unspoiled to this day. Who was at the top of the hierarchy of this unspoiled cosmos—God?

M: No, the most powerful person in the valley was the man with the biggest farm. Then there was the parish priest, a venerable old gentleman who we all respected. My father was the senior teacher and headmaster of the school in the valley.

H: Your father also bred rabbits. Why was this?

M: We were nine children, and my father needed the extra income. My mother used to shear the rabbits and sell the angora wool. We butchered a few as well, but we kept the fifty or sixty rabbits mainly for their wool, angora wool—very valuable wool.

H: And you also had a chicken farm.

M: We ended up with thousands of chickens. We supplied chicks and pullets all over South Tyrol. All of us children had to lend a hand. I started working in the henhouse at the age of six.

H: How many hours did you have to work?

M: In the summer six, seven, eight hours.

H: A day?

M: Yes. My father wouldn't have been able to support the family with his teacher's salary alone. He also had a couple of ulterior motives when he lumbered us with work. Firstly, he didn't want us, as the children of the village teacher, to have a privileged position in the valley. All the kids in the valley had to work. Just messing around and playing or being able to afford to do nothing at all was regarded as immoral. Farmers' children had to work in the stables or cowsheds, look after the animals, cut corn, and bring in the hay. So we had to pitch in and help, too—with the chickens. Secondly, by giving us work to do, he kept us off the street and away from any possible vices.

H: In the football film *The Miracle of Bern*, a rabbit is butchered for the Sunday roast. When the young boy finds out, he goes to pieces. Did you have a similar relationship with animals, or did you just regard them as a useful commodity?
M: On Saturdays, when I was about ten years old, I used to kill and pluck as many as fifty chickens on my own.

The mother of Frau Degani at the Hotel Kabis also kept chickens, and she used to come and fetch us when she was making chicken for her guests. "Boys," she would say, "come and help me pluck some chickens." Killing them and plucking them, it was like doing homework. We did it for the parish priest as well.

H: How do you kill a chicken? Chop its head off?
M: We had our own method. Father taught us how to do it. You hit the chicken on the temple, just above the right eye, with a big pair of scissors, dressmaking scissors. That stuns it. All very simple. You clamp the chicken under your left arm so it can't wriggle—I can still do it today, as easy as picking up a pencil. The chicken isn't frightened; it's all quite normal. I hold its head in my hand and hit it hard on the temple with the big scissors. The chicken is unconscious for a few moments. It doesn't feel a thing. I hold the beak open with two fingers and slice through the arteries in the roof of the mouth. It only takes one cut, as I can feel exactly where the soft part of the palate is. The chicken bleeds to death. The blood just runs out and there's not much wriggling about. The chicken is still stunned. It moves a bit at the end, shakes itself, and then it's over. But if you chop a chicken's head off with an ax, it runs around without its head.

H: You view such methods with contempt?
M: Yes, because it's unprofessional. It's horrible. I can't watch that. Because it isn't skillful.

H: It seems you were a perfectionist even then. How long does it take you to pluck a chicken?

M: If the chicken is still warm, ten to fifteen minutes. I know exactly which parts I have to take care with so the skin doesn't tear. Some days I'm faster than others.

H: After the Second World War, South Tyrol still belonged to Italy, but most of the South Tyroleans hated the Italians and felt like they were under foreign rule. Were you aware of this sentiment as a child?

M: In the 1950s there was a strong anti-Italian feeling. That was understandable. In the 1930s, during Italian fascism, we had been beaten down and Italianized. In Rome they used to say, "Within fifty years everyone in the country will be speaking Italian; then there won't be any German South Tyroleans left!" Even the schools were Italian-only. Try to imagine it: you're twelve years old and from one day to the next you have to attend an Italian school, and the Italian teacher can't speak a word of German.

There was terrible disruption. People had no idea what was happening to them. And they felt so badly treated that in 1939, when the Hitler-Mussolini "Option" was agreed to, 86 percent of all South Tyroleans chose to leave South Tyrol and become citizens of "Greater Germany." Nearly all the workers, many, many farmers—who had their homes here, their possessions, their farms, people for whom home was everything—they voted for Germany.

H: It sounds like you have a degree of sympathy now for this "home in the empire" sentiment.

M: On the contrary. One of the first scandals I stirred up in South Tyrol was when I claimed that this amounted to a betrayal of our homeland. I feel sympathetic toward the ordinary people, but I have no sympathy with the political leaders of the time.

Most of the people told themselves it was better to leave the place that had been their homeland for a thousand years than to stay there under Italian rule. Their true homeland, and the focus of their German-ness, became Hitler's Germany.

Ultimately, they also wanted to please the Führer by aligning themselves 100 percent if possible with Germany. They hoped that by doing that, the emigration would somehow not take place, even though Hitler had stated in *Mein Kampf* that he didn't give a damn about South Tyrol. The Berlin-Rome axis was far more important to him than German-speaking South Tyrol. Such cynicism! And so much gullibility!

H: Was your father also one of those who voted for emigration?
M: Yes. He thought we'd be resettled in the Carpathians, the Crimea, or wherever. The whole resettlement thing was actually pretty vague. People were promised a lot, but the promises weren't kept. The "optants," the ones who chose to emigrate, were housed in temporary settlements. The farms were all surveyed, and the people were told that they would get a similar farm to the one they had here. There was a lot of talk—public talk, too—and a lot of propaganda. The winegrowers would go to Crimea, the hill farmers probably to the Carpathians.

H: So your father was actually in favor of doing the same thing to others that he himself had suffered—taking away their land, eradicating their culture, oppressing the people.
M: In the summer of '39, several leading South Tyrolean politicians quietly went to Berlin to seek an audience with the Führer. They didn't get an appointment, but they did get some information. They were seen by Himmler. They asked him, "What will happen to us if we vote for the Führer?" Himmler is said to have told them, "You'll go to the Carpathians or Crimea, as a united people."

Ten years after the war, my father gave me a children's book about bear hunting in the Carpathians. "Read that," he said. "That's where we'd be today if things had turned out differently. It's very interesting. There are mountains there as well." So the Carpathians are the mountains we'd be living in today if history had been different.

The South Tyroleans like rattling on about their love for their homeland as if it's their greatest asset, their great strength. But back then, in 1939, nearly all of them went away. I still can't figure this behavior out. And I find the concept of "homeland" a bit suspect.

H: Was there a noticeable feeling of German nationalism at home?

M: Yes. You can see it in our names. I was christened Reinhold because the name can't be Italianized. My oldest brother is called Helmut; his name can't be Italianized either. The same goes for my sister's name, Waltraud and Günther, Erich, Siegfried, Hubert, Hansjörg, Werner, and so on. We didn't have a Josef, as the Italian fascists would have made Giuseppe out of it.

H: Did you ever speak to your father about the madness that was Hitler's Germany?

M: I tried, but it was just dismissed out of hand.

H: In what way was it dismissed?

M: We older children naturally had questions, but Father said nothing and Mother told us to stop asking. Then, in a quiet moment, she told us, "You have to understand that you must never talk about it. He can't cope with it. You mustn't speak about the war, the Nazi time, or the persecution of the Jews."

H: When did you realize that things like the Holocaust had happened?

M: I was maybe fifteen. Before that, I didn't even know that Jews had been killed. They still sang those old songs from the Wehrmacht time in the village—songs like "Am Bahnhof von Jerusalem, da kann man Juden sehen" ["At the railway station in Jerusalem, that's where you see the Jews"]. I heard them at the inn as a child, and when I got home I asked what they were singing about. My father told me to "stop that nonsense!" I couldn't understand why he was annoyed—it was only a song they were singing. I didn't even know what Jews were. I thought, "What's wrong with him?"

H: Why do you think your father wanted to shut the door so firmly on his past? Could it be that all that German nationalist behavior was embarrassing for him? Perhaps he also felt bitter about it?

M: I think he felt cheated of his youth. And although he never admitted it, I think he felt that the war had caused him to abandon his ideals. Before the war he'd attended a seminary and he'd been climbing, and then, at twenty-eight years old, he'd come home—empty, disappointed, without hope.

He became a teacher because he needed a job. It was only later that it became his profession. This option was open to anyone who had attended a grammar school. They needed German-speaking teachers. The German school wasn't abolished after the war, even though we remained in Italy. My father was a self-taught teacher, brilliant at explaining things, but he certainly wasn't a good educator.

H: How did your mother manage to put up with this gloomy, strict man?
M: I don't really know. She had to stay with him, I suppose.

H: In many respects she seems to have been the exact opposite of your father.
M: She was called Maria—she had dark hair and the radiance of a Madonna figure.

H: What color were her eyes?
M: Bluish? I can't remember. Funny that I can't remember that now. Anyway, she didn't have the same blue eyes as us children.

H: What comes to mind when you think of your mother?
M: She was always there, and she solved all the problems. She was the calming influence on the family. She was there for everyone. She had an altruism that came naturally to her. She also had unlimited patience. I still don't know how she found the time to do everything.

H: Can you remember ever being shouted at?
M: No. No, Mother seldom shouted at us.

H: "No" or seldom? Can't you decide?
M: She was gentle. She was the counterbalance to our stern, unhappy father. Whenever one of us boys was in trouble—our sister was very well behaved—Mother always managed to remedy the situation. My brother Hubert was kicked out of grammar school for reading a Heinrich Heine story out loud in the dormitory. Heine's journey through South Tyrol to Italy, with his

nasty description of Brixen. It's actually a beautiful piece of writing and very apt. Our father ranted and raved about it, and then he said, "Okay, if the lad is so stupid, he can forget about school and go and do an apprenticeship or get work on a farm." But Mother went out the next day to find a new school for him. She went to Meran and found a headmaster—an ex-comrade and school friend of my father's, as it happens. Father didn't go with her. She got Hubert a place at that school in Meran. Hubert took it and later went on to study at university. Now he's a very successful doctor.

H: Did you ever need your mother's help?

M: Yes, I've always been a bit of a revolutionary. And I've invariably had problems with people trying to lay down the law with me. Like my father. I was the first to rebel against him. This often caused arguments. But somehow my mother always managed to settle our disputes. Otherwise, my father would have killed me.

H: Can you give me an example?

M: There were the early times on the poultry farm when I went off skiing before I'd finished the work I'd been told to do. Then there was my passion for climbing. That started when I was five, when I climbed Sass Rigais for the first time. My father encouraged me at first, but he soon put the brakes on and started to limit my excursions. Probably because he realized that I was hugely enthusiastic about climbing.

We weren't allowed to get passionate about anything. That was frowned upon. We could do anything we wanted, but only if we behaved in a way that was acceptable to the good people of the village. One time we were in church—us older children—and when the priest started spouting some nonsense or other, we stood up during the sermon and walked out. Down the aisle wearing nailed boots. Out of protest. To show people we weren't going to stand for any nonsense.

H: What "nonsense" were you protesting against?

M: I can't remember. There was terrible trouble at home. But Mother said, "Go on with you. Leave the boys alone; it's not a problem." "What? Their so-called protest has insulted the entire village community, and I am the village teacher. The parish priest is an institution."

H: You have said you are actually in favor of a matriarchy where family matters are concerned.

M: That's right. Our mother set an example of matriarchy in the way she organized the family, the way she solved problems, the ability she had to always find a way out. It worked. It was successful. And because I also instinctively knew that patriarchy would in our case have led to catastrophe, I am all for matriarchy. I'm an experienced man in that respect.

These days, Sabine, the woman I live with, is the one who makes the decisions at home. Well, we make them together, but she has the last word. That's the way it should be.

H: What part did God play in your childhood?

M: No more than the parish priest. A habit.

H: Did you think of him sitting up there on a cloud, or what was your idea of God?

M: I didn't take religious studies too seriously really. But I didn't come over all revolutionary and proclaim that God is dead either.

H: God is present everywhere in Villnöss. You see those little Madonna pictures all over the place, you have to go to church on Sundays—you must have formed some kind of idea, surely?

M: Yes, those pictures were everywhere. And we were inculcated by outside forces. But in spite of all that, I never really formed a mental image of God. It wasn't a revolt against God as such. When I walked out during the sermon and started skipping church on Sundays, it was just a personal protest against the apparatus of the church.

People didn't go to church because they liked attending mass. They went to church because that's what you did. Everyone went to church. It was a habit. It was unimaginable that anyone would stop going to church. We were probably the first in the valley to say we were going climbing on Sundays instead. Mother said, "You don't have to go to church." So we set off at five, long before the early mass, and walked from our house up to the edge of the forest and went climbing on the Geisler.

H: Did you have your own room?

M: No! For a long time there were six of us boys sleeping in one room.

H: That sounds like murder.

M: In bunk beds. Our sister had a tiny little room of her own. Our parents had their room, and my father had a kind of study, where he corrected piles of homework. There was a kitchen-cum-living room. No bathroom.

H: When there are six of you in one room, is there any such thing as personal possessions?

M: I owned some toys, an ax, and a pair of skis little wooden skis, my first big Christmas surprise.

H: And what else did you own?

M: Two pairs of shoes, two pairs of trousers, one pullover. From my older brother. We were always neatly dressed, though.

H: What did you have to eat and drink?

M: Our food was a combination of Italian pasta dishes and typical South Tyrolean fare. There was a fixed weekly meal plan: Monday, dumplings; Tuesday, something else; and Wednesday, something else again. But the same thing every week. Year after year, for decades on end. Bacon dumplings or something fried or roasted. Meat once a week. Chicken from our own chicken coops or rabbit. We drank water. Milk and bread and jam for breakfast. Never wine or beer.

H: And fruit?

M: We had a few apple trees of our own. We had a tiny piece of land with a few things growing on it: cherries, plums, black currants. We also had permits to pick fruit and collect wood. We went up to the woods to pick cranberries and raspberries in early summer. We often picked 30 or 40 kilos of berries, which were then boiled up to make jam. We also picked chanterelles. And we collected firewood from the forest so we could heat the house in winter. We cut ferns for the chicken coops as well. They were a good remedy for lice.

H: Did living in a narrow valley give you a feeling of confinement or one of security?

M: It always seemed to me like I'd never get out. The valley was my whole world. The clouds came in on one side of the valley and then disappeared on the other side ten minutes later. What lay beyond didn't exist. Maybe my wanderlust has something to do with my childhood memories.

H: Did this feeling of confinement scare you?

M: I was never scared of enclosed spaces as a child. Quite the opposite. I wasn't scared of going hungry either, or worried about making ends meet. Somehow everyone in the valley seemed to manage in one way or another. The worries and fears came later, imposed on me by outside influences. "If you don't finish your studies, you won't get a job. If you don't do as you're told, you won't achieve anything. We haven't got a farm, so you can't just stay at home and look after your cows."

H: Were you curious about the world beyond Villnöss?

M: I certainly was. My first big mountain route on Sass Rigais shook me up a bit—not because it was so long and strenuous but because I was able to look out beyond our valley. There was a valley, then another valley, and then more mountains far away in the distance. I couldn't see what lay beyond that. But the world suddenly became bigger. My universe expanded. My curiosity was awakened. What lay beyond became the big question.

H: What did you think lay beyond your valley?

M: Well, since I'd seen that beyond our valley there was another valley and beyond that another one, we walked over the mountains. Over the Geisler at first, and down to the alpine pastures on the other side. There were marmots over there, and streams and lakes. It was a more mysterious world than our own local mountains.

H: And the desire to explore grew?

M: The desire to strike out and see what lay beyond, yes.

H: What did the mountains mean for the other folk in Villnöss? Did they also enjoy climbing dangerous lumps of rock? Were the mountains also a symbol

of freedom for them? Or did they believe that up there was where the evil spirits lived?

M: You went into the forest to fetch wood and up to the high pastures to fetch hay, but no one went any farther than that. Apart from a few poachers maybe. You went as far as you had to go to get what you needed: wood, hay, or game. Only stupid people went any farther than that.

H: Why stupid?

M: Stupid because they didn't understand the world. People who didn't have anything sensible to do at home.

H: Stupid people with too much time on their hands?

M: Too much time, too little work, no land or possessions, or people who just didn't know what to do with their lives. Sensible people didn't go any higher than the high pastures. Well, apart from the hunters, who might go up into the mountains to shoot a chamois from time to time.

H: The mountains were a threatening place with their avalanches and rockfall.

M: But they never came down into the valley. The land that the people of Villnöss worked was safe from avalanches. They thought that everything up to about 2300 meters could be used—for grazing, pasture, or woodland. Anything above that was deemed out-of-bounds; the world up there was taboo. The cliffs up there were not sacred or exalted, just worthless. The world up there was of no use to them.

H: In the Middle Ages people were terrified of the mountains. There are stories of people being carried over mountains in sedan chairs with the curtains held shut, because the sight of the dreadful rocks up there was more than they could bear.

M: It wasn't like that when I was a youngster. The mountains, and what lay above and beyond them, were something the local people just weren't interested in.

H: How did you imagine the world beyond the mountains looked?

M: I thought it must have been similar to our world. We didn't have a television when I was a child, and I never listened to the radio or saw a newspaper. My parents never really went anywhere bigger than Bozen. I think they'd been to Dresden and Venice as well. They certainly talked about it—like fairy tales from times past—but I couldn't really imagine what it might be like there.

I took geography at school: South Tyrol, Italy, and Rome. We looked at a few pictures of Rome. But I never imagined going there. I never dreamed of going to Rome, to that magnificent capital of the Roman Empire.

H: Saint Peter's Square left you cold?
M: I dreamed of a summer climbing in some gorge or other, getting to the top of this big face we'd gotten scared on and backed off the previous year.

H: When was the first time you went climbing?
M: When I was five years old. There were four of us: my father; my mother; my elder brother, Helmut; and me. We went up to the Gschmagenhart Alm first. We slept in the hay, fetched water and firewood, and at the end of that first little stay up on the Alm we boys were allowed to climb Sass Rigais. With our parents. My father taught us the basic climbing holds and moves on some of the little crags and boulders up there. And we had a rope with us, an old hemp rope my father had kept since he was a boy.

H: Did you feel safe with that rope?
M: There were some people ahead of us, and a few little rocks came down. I remember that the rockfall made quite a big impression on me. My father shouted up, "Careful! We're climbing right underneath you!" It was immediately clear to me that if you got hit on the head by a rock you'd have more than just a bruise.

We climbed ahead of Father, without a rope. The climbing was easy. As I remember it, the first and only scary situation was on the summit, looking down the 600-meter vertical drop of the north face. I guess it was that route that stirred up my passion for climbing. I didn't just have the ability and the endurance required; I was at the age when I was starting to create my own world. My elder brother coped well with the climb, too.

After that, we went back to school and spent the whole winter talking about the mountains and about summer. When I was about ten or twelve years old, I started looking for new routes, and the peaks at the end of the valley became my place of refuge. By then, my brother Helmut had become the star student at school.

H: What was the decisive factor that awoke your passion for climbing?

M: I can't really say. Maybe it was the praise I got from my parents, especially from my father, who'd taken his little boy climbing and he'd actually done pretty well. Or maybe it was just the chance to let off steam, to break free, to climb a long route. I don't know.

H: Wasn't the climbing simply about increasing self-confidence and establishing a personality?

M: I wouldn't put it that way. The next step was that my younger brother Günther and I went climbing on our own on the Geisler peaks. On the north side. We didn't do any of the hard routes, but even so, the fact that we were allowed to go at all was a miracle.

With hindsight, I admire my father for his generosity in that respect. We didn't know how hard the north face of the Kleine Fermeda was. We didn't even know where the route went. We just set off climbing, worked out where to go, and ended up on the summit. It made us feel strong, for sure, but I don't know about the personality thing.

H: What was your first difficult rock climb?

M: I did my first real route with my father. On the east face of the Kleine Fermeda. It wasn't actually that steep, but it felt vertical to me at the time. I must have been about twelve years old.

H: What grade was it?

M: Grade III

H: What does grade I look like?

M: Grade I is when a good climber needs to use his hands to do the moves; otherwise, he'd lose his balance and fall off. Grade II is moderately difficult. The normal route on the Grosse Fermeda is grade II, maybe even II to III. And grade III can be described as difficult for a competent climber. That means you need experience, strength, and endurance. You have to know how to climb, and you need to have basic routefinding skills.

Climbing is also about power-to-weight ratios. That means a child won't necessarily find climbing itself difficult. A child also has small hands, so he can hang on to small holds. The disadvantage is that he won't be able to reach as far. If the next handhold is out of reach, it's out of reach.

H: How would you describe the climb?

M: There's quite a difficult approach, then chimneys, grooves, and gullies. It's not exposed, but some sections are smooth and polished.

H: What do you mean by "exposed"?

M: It basically means that there are big drops all around. But between the Grosse Fermeda and the Kleine Fermeda, there's a series of narrow channels, waterworn grooves, and gullies, which become watercourses in a thunderstorm. That's what you climb. After that, the route takes a system of chimneys splitting the vertical face.

H: And how do you climb these chimneys?

M: By "stemming" or "bridging"—bridging the gap with your feet, legs apart, using opposing pressure on the sides of the chimney. A wide crack or chimney like the ones on the Fermeda will sometimes get narrower for a bit and then open out again. The next part of the climb is less steep. Then it steepens up again toward the top. The actual steep section is maybe 200 meters high. And vertical.

Father knew this final summit wall from an earlier ascent, so he led the first bit. After 40 meters, one rope length, he tied the rope to a rock spike and brought me up. On the last third of the wall, he let me lead a pitch, a full rope length. He deserves great respect for that, as it could have all gone badly wrong. If I'd fallen off, I'd have been dead. In those days, climbs were not as well protected as they are now, and I had no gear in.

H: That would have been a long fall.

M: There were hardly any in situ pegs on that wall. Just one or two old ones. We clipped a carabiner into them and clipped the old hemp rope into the carabiner so it ran through it freely. Whether it would have held a fall is questionable. That kind of thing was out of my control. I had no idea. But I was allowed to lead a pitch for the first time.

The question was where to go—in general and in detail. I obviously got the hang of routefinding pretty quickly. But I couldn't afford to fall off. Perhaps my father wanted to show me what leading was all about: communicating with your partner, taking responsibility, being alert. And above all, being careful. My father did, of course, tell me before I set off that I wasn't to take any risks and that I mustn't fall, as that would be extremely dangerous.

After that route I was allowed to do similar climbs with my younger brother. On our own initiative. Hard to imagine parents letting a couple of lads that age out to go climbing in the mountains on their own.

H: Your father had seen that you could do it, and he consciously took a chance and trusted you.
M: That's right.

H: And trusted you to come home alive.
M: Yes! Coming home alive was the deal. So there we were, two young lads starting to climb on our own. We said, "We're going to do the north face of the Kleine Fermeda tomorrow!"—a climb that Father had never done. All he said was, "Just be careful," and he asked us if we knew where it started and finished.

H: How do you explain that? You portrayed him earlier as a controlling, almost pedantic man, yet now it seems he was holding the door wide open for you.
M: It was certainly never our father's intention to encourage us to embark on a climbing career. But he didn't want to forbid us doing the things he'd dreamed of doing before the war. The routes he'd done when he was twenty, we were allowed to climb when we were twelve. There was never any feeling of rivalry.

H: Maybe he was acknowledging the fact that his own dreams had just ebbed away?
M: It might have been that. He'd accepted the fact that his climbing life had been interrupted and probably told himself that he wasn't going to prevent his boys from achieving their own dreams, that although we were young,

we were good climbers. I think he knew that I was a better climber than he was, even though I was naive.

Father and I also climbed the east face of the Grosse Fermeda together, and I led the whole route. But Father was still the one who called the shots: "Go left! Careful—rock!" He'd done all these routes before.

What he and his student friends had not managed to do before the war was the north face of Sass Rigais. So he went on it with me. I can well remember my mother telling him, "No, that face is too big and too dangerous for the child. You can't do that." He went on the route with me anyway. We climbed up a little way. He went up and back down again, and then said we should try over to the left. Up again, down again. We never really got going. He kept climbing up and then reversing back down. He was scared. We finally backed off the route, and Father said he couldn't work out where it went.

So what did Günther and I do? I was sixteen at the time, I think, and my brother was fourteen. We decided to climb the route that my father hadn't been able to do. Not to show him what a useless climber he was but to make a statement about what we wanted to do. We were moving into another dimension of climbing, one that my father did not know. By then, we had a few pegs and a helmet, a nylon rope, and a peg hammer—the essential items of gear.

H: Your father gave you that first helmet for Christmas, didn't he?
M: Yes. The peg hammer was a present from him, too.

H: What did the hammer look like?
M: It was huge. A big, heavy whopper of a thing, made by the village black-smith. It was his, a relic from the 1930s.

H: How heavy was it?
M: Twice as heavy as the peg hammers you could buy in sports shops back then. It had a longer shaft, too, a good 30 centimeters long, just like the hammers that blacksmiths used to make. It didn't have a pick on it—there was no point at the end to chip away at things with. It was expressly made for hammering in pitons. It was a decent hammer, great for rock routes, and much better than those little hammers you could buy

in sports shops. It did the job well, it felt good in your hand, and the pegs went in quicker. I had it with me on some of my boldest first ascents in the Dolomites.

H: How can you tell whether a piton is in a secure placement?
M: By the sound it makes. A well-placed peg "sings" as it goes in. If it goes in too quickly, forget it. If it makes a dull, feeble noise, it won't hold. But if it sings as it goes in, it will hold. A few thousand kilos and more.

H: Did you like bashing in pitons?
M: I got good at it. Out of necessity. My enlightenment came at the age of twenty. I was one of the first climbers to do things in a different way from all the others. I didn't just hammer in a quick peg when the climbing got tough. I climbed, and when I noticed that here was the perfect crack and I could stand in balance, I placed a peg. For protection. It took no time at all, then I kept on climbing. When I came to a tricky section, I didn't have to look around for a crack, a crack that might not actually be there. I was already protected. Standing in an exposed position and placing a peg means losing a lot of strength. I was able to climb quickly on the hard bits because I'd already placed some protection.

So I didn't climb like all the others; I did the exact opposite. Previously, when things started getting difficult and they couldn't do the moves, people had always just bashed in a lifesaving piton somewhere or other—which as often as not wouldn't have held a fall. I placed the lifesaving piton where I didn't actually need it but where I could stand easily and place it well. That meant I didn't lose much time or strength.

Later on, when I was in my early twenties, this tactic gave me an edge over some of my peers. We did new routes that appeared to be unrepeatable, because there were no pegs on the crux sections. The climbers who did the second ascents often took three times as long as us. We weren't any better; we just had a different approach to the rock.

H: What gave you the self-confidence to try these new techniques?
M: Throughout my life I've always had respect for the past, by and large. I've learned from others, and I always accepted a subordinate role to start out. Don't forget, I came from a little valley. When I went out with famous

climbers for the first time, I thought, "This is great!" I was just a little kid with a vague idea of climbing. Then I realized that there were some things they really could do much better than me. So I watched and learned. I learned a lot from them.

H: For example?
M: Belaying, rappeling, pacing myself. But I also noticed that they weren't as good as me at certain things. What I was better at was routefinding.

Since the age of eighteen, I've never met a climber who could spot the line of a route better than my brother Günther or I. Routefinding was something we'd learned to do when we were children, in the years we'd spent climbing the chimneys and grooves on the Geisler. We'd done countless routes without reading up on them in the guidebook first. We just knew that the route went this way or that, which line was the best one to take, or the only one. When we did the north face of Pelmo, we knew you had to go right and after that it's okay. Other climbers never developed that instinct, especially the city boys.

H: You didn't make yourselves very popular, did you?
M: No, but that wasn't our aim.

H: Were you a know-it-all?
M: Not at all, no. How else was I going to learn? When I started out, I bowed to the experience of others. But as soon as I realized that they really couldn't do it, I made the decisions. Just like that. And the others immediately went along with it. Without hesitation. There was never any rivalry between my father and me. Or between my brother and me. When we climbed together, it was, "I'll lead, you second." That was never an issue. I was the older brother. We never argued about it. Whoever led the route assumed more responsibility. That was obvious. I was bigger, and I was more experienced.

H: In your early childhood, you and Günther didn't really get along too well. Why was that?
M: My brother Günther looked exactly like my father: slim, wiry, with a

hooked nose and black hair, black as coal. I looked more like my mother. Günther was stubborn and strong-willed; I was an obstreperous child. I used to lose my temper and really fly off the handle. My brother would just dig his heels in and say, "I'm not putting up with that. I'm not going to say any more now, but I'm not putting up with it." We grew closer through climbing, through the secret world above the clouds that we had discovered for ourselves. As contrasting characters, we complemented each other well. Together, Günther and I were unbeatable.

H: Did you used to consciously avoid Günther before that?
M: We went skiing together, and we had to work together in the henhouse.

H: Did you ever come to blows?
M: I can't remember ever coming to blows with Günther. We had fights with other kids in the village, and in school. Frequently, in fact. That was normal behavior back then. For a short while, just before the end of primary school—so I would have been about twelve—I was quite a feared scrapper.

H: Were you the toughest kid in school?
M: There were a few lads who were bigger and much stronger than me. I never took them on. They were considered to be the tough guys, so you kept your distance. In any case, they wouldn't have touched me, because I was a lot younger; I was out of their league, so to speak. But in my own class, I was known as a fearsome brawler.

It wasn't that I was particularly skillful or strong, but even then I had plenty of energy and endurance, and I could be very aggressive. Why? I don't know. I can still have these aggressive outbursts even today. I have no idea where it comes from. But I can get agitated very quickly. It's part playful, part serious. But when faced with injustice, aggression, or deceit, I can get dangerous. And when that happens, I can frighten twenty men. Unfortunately. It shouldn't be necessary.

H: Did you used to hit people, or was it mainly just screaming and shouting?
M: We never used our fists or threw stones; it was just a matter of showing the other boy who was stronger.

H: Until one of you gave up?

M: Until one of us was lying on the floor, or until one of us said, "I give up!"

H: There's that well-known incident when you were twelve and you found Günther cowering in the dog kennel outside your house, whimpering because your father had thrashed him so hard with the dog whip that he couldn't walk. That was when you realized that Günther was just as much a victim of your father as you yourself were, and you became friends. You went on to become one of the most legendary climbing partnerships in the history of alpinism. But first, a very simple question: what is a dog whip?

M: It was like a rubber baton. Not a baton like the police have, but a length of rubber tubing that my father used to hit the dog with until it became obedient.

H: What had Günther done wrong?

M: Something trivial. It's not important. I wasn't there. For all intents and purposes, that situation was just a fact of life, an everyday occurrence in the postwar era.

H: But that is a terrible story.

M: Yes, it's a terrible story.

H: When a child is so mistreated that he can't walk.

M: It didn't just happen to us. Children were beaten in our valley, in South Tyrol, and throughout the German-speaking world. Fortunately, not everywhere.

H: Beating as an education method?

M: That was the normal method of education in our valley. What went on beyond the valley, I can't say. I wasn't there to see it at the time.

H: What did you get beaten for?

M: For everything. For not feeding the chickens at the right time, or for fighting in the house. As small children, we got slapped for shouting or just playing. The house was small. With so many children around, the only way

A young Reinhold Messner on the Cinque Torri in the Dolomites (1965)

Father could keep order was with his cane. It was a case of spare the rod and spoil the child, I suppose. So yes, he hit us.

H: Was it systematic punishment, or did he hit you in the heat of the moment?
M: Both. A deliberate parenting method and an impulse reaction. Maybe it was a kind of escape for our father from the despair he felt at the stalemate situation he found himself in after the war.

H: Stalemate situation?
M: After the war, my father had no opportunity to shape his own life, to develop as a person. He had no alternatives. Previously, he'd been a man with ideals, vision, a future. He'd been a good climber, he'd studied, he was a South Tyrolean who had imagined a life for himself.

H: He stood with his back to the wall and never climbed it.
M: All his life he stood with his back to the wall. It was all to do with the war, with subservience and conformity, and responsibility for a large family. I mean, those young soldiers had seen dreadful things, and couldn't come to terms with it. He had nothing left, no hope, just a kind of word-less despair. And always with his back to the wall.

We all know the experiences that our fathers and grandfathers had when they were torn from their worlds and plunged into terrible disaster, even if it's just from literature. We can't comprehend now how almost all of them became co-perpetrators. And then they came home and maintained that it wasn't them. But it was them, and they know it. In *Morbus Kitahara*, Christoph Ransmayr explains this so impressively that it scares me.

H: Did you ever hit back?
M: No, never. But at some point I think my father sensed that he couldn't hit me anymore. When I was thirteen or fourteen, I was stronger than him. He knew how far he could go.

H: When you were twelve years old, what kind of job did you think you'd end up doing?
M: Back then, I had no career aspirations, no job plans, no real idea of what

to do with my life. I only ended up at the Geometerschule in Bozen doing a kind of civil engineering program because I was good at math. Technical and scientific subjects suited me. But I never had any desire to become an engineer. An architect maybe.

H: Lots of boys want to be a train engineer or an astronaut.
M: I never wanted to be a mountaineer, if that's what you mean. I couldn't, because it wasn't a proper job. The way my life turned out was inconceivable at the time. What I dreamed of was unrealistic: to keep on climbing even if the world came to an end. From the age of sixteen to eighteen, I actually imagined myself leading the life I went on to lead. In spite of all the opposition and the initial self-doubt.

H: And that was?
M: I wanted to be an adventurer, but not an explorer in the established sense.

H: A gypsy?
M: Yes.

H: A modern-day gypsy?
M: I always wanted to visit the last great wildernesses, places where others couldn't get to so easily.

H: Did you have any role models?
M: No, there were no role models, just characters who were like me.

H: Was this a kind of Jack London fantasy? Did you read books like that?
M: No, and I never read Karl May either. But I wanted to go into the wilderness. We trained our dog to pull a sled. I just wanted to head off into the wilderness, to climb cliffs, to roam the countryside; that's all I wanted to do.

H: And how were you going to feed yourself?
M: The question never occurred to me. In any case it was obvious that it wasn't going to happen. No one could make a living from conquering the useless.

H: Back to the story with your brother. What was it that brought you two together on that afternoon, when he sat helpless in the dog kennel, unable to walk? Did it ever dawn on you that you and he would go on to create your own world in the mountains?

M: We became sworn allies; that much was clear. They could boss us around and compel us to do things, but they couldn't break us. We'd build their chicken coops for them, but not forever!

Maybe our father had a guilty conscience, as he did let us go out climbing on our own later. When I think about it, I was eighteen, my brother was sixteen, and he let us go off to do the north face of Monte Pelmo, a gigantic face, 800 meters high, vertical. A really big wall. And we got caught in a storm as well. Father lent us his Lambretta scooter to get there. We started up the route at five o'clock in the morning the next day, a route that had been climbed maybe a couple of dozen times, by the best climbers in the Alps.

H: You were caught in a storm?

M: The climbing was hard at first. The rock was smooth and compact, the climbing harder than we'd expected. After that, it all went smoothly, very quickly. We were in good shape and had everything under control. Then, around midday, the storm came.

H: You were, what, 400 meters up?

M: Yes, fresh air below us, and above us lightning and hailstones. Catastrophic!

H: Were you scared? Where were you?

M: On the easiest part of the face. We sat under an overhang at first, where we were safe from rockfall. There were rocks coming down continuously. It became clear that we couldn't get back down the face, as the lower you went the more falling rocks there were. One of them would have hit us for sure. So it was onward and upward.

Later on, the storm stopped and it even brightened up. Then it got very cold. The rock was wet. And then, on the last quarter of the route, the next storm came. Snow, hail, lightning. The rock was white, verglassed. It was bad. That's what real adventures are like. Either you survive or you're dead.

H: Was that clear to you at the time?

M: For sure. In situations like that, you know that it's all about surviving, but you don't say anything. There's no going back; the only way out is up. Your instinct kicks in, and you do the right thing. But if you're unlucky, one mistake and it's over.

H: Surely there's a contradiction there? Do the right thing and you won't die, make a mistake and you might?

M: No, it only seems like a contradiction. Maybe it's a mistake to climb the mountain at all, but once you are in danger, the survival instinct mobilizes all your strength, the seventh sense, fear and courage. It just happens. You instinctively behave in the right way. There is no longer any doubt. Doing the wrong thing just doesn't happen.

H: You can still get killed, though.

M: Yes, you can get killed in the mountains.

H: So you climbed on through the storm with wet, numb fingers. Was that the first time in your life that you felt you might die?

M: That was the first time. I knew we were on one of the hardest alpine faces in the middle of a storm. Only the very best survive that kind of thing. I haven't been in that many life-threatening situations in my life. A hundred times maybe.

H: Once more: this was the first time you'd experienced a life-threatening scenario?

M: Yes. Either you make it or you're dead.

H: You had a real crisis to deal with there. Afterward you must surely have asked yourself, "Do I keep on climbing or pack it in, go to technical college, and become an engineer"?

M: A reasonable question, but the thought never crossed my mind. And at the time, I had only one thought: Let's get out of here!

We got to the top in the dark—hypothermic, clothes frozen stiff. There was thunder and lightning, and it was snowing. We descended on the south side. The fact that we found the descent route at all was entirely luck. The only light was from the flashes of lightning. It was pure instinct that got us down to the Ball Ledge, and from there we made it down to the forest.

H: Did you not think afterward, "What on earth have we just done?"

M: It was different. We thought, "It can't get any worse than that. If we've survived that, we can handle anything." We told ourselves we'd never be that unlucky again, to be caught in a storm on such a big face, with rock-fall, snow, lightning, freezing cold.

H: Did you have a feeling of immortality afterward?

M: Yes, a bit. There was a kind of naive "it won't get us" feeling. It wasn't that we were any better than the others, but we'd just been through hell together; it couldn't get any worse than that. We were up for something harder next time. We'd do it, too, even if the weather didn't play ball. We felt like we'd now joined the elite circle of extreme mountaineers, and not just because of our climbing skills.

H: How big was the circle of extreme mountaineers back in 1964?

M: In South Tyrol, about twenty people; in North Tyrol, maybe twenty-five; in East Tyrol, three or four. We all knew each other, by name at least.

H: And in the entire alpine region?

M: A few hundred. There are fewer climbers of that type around nowadays. But there are a hundred thousand or more in Germany alone who climb on indoor climbing walls.

H: And what do the extreme mountaineers do?

M: They do things that other people regard as unreasonable, irrational. Life-threatening things. Things that require skill, endurance, and disci-pline. You have to be tough to sleep out in the open at minus twenty degrees. You have to be able to do without food or drink for a couple of days if there isn't any. You bear responsibility for yourself and for your team. You have to do everything yourself, you and you alone. You operate in a world where humans do not belong, a place where it seems totally irrational to go.

We extreme mountaineers go to hell of our own free will, and our response to the critics is, "Leave me in peace; I've made my decision, and I want to give it a try." And when we return, we return to a like-minded fellowship, a kind of clique, I suppose, with its own jargon, its own language. Wannabe climbers can never be part of it, as membership can't be bought and you can't talk your way in. You've got to walk the walk.

H: A kind of subculture?
M: Yes, the extreme climbing scene has its own culture.

H: The American journalist Tom Wolfe described a similar type of culture among astronauts and test pilots, a culture with its own moral code. Does the same thing apply here?
M: Of course. We would never talk to each other about morals, though, because we all know that morals are a purely bourgeois affair.

When you're in trouble up there with no hope of rescue, you'd do anything to save your climbing partner—carry him, drag him, shout at him, anything. If you've got two mouthfuls of water left, you give your partner half or, if it's a matter of saving his life, all of it, the whole bottle. It's not something you ever talk about; it's taken for granted. And if someone comes along and says you have to share the water, well, that's just idle talk. Out there, beyond the human world, it's all just taken for granted and we don't speak about it.

Then you get the self-important ones, who complain that someone or other abandoned his partner—it's stupidity. Or pomposity. This moralizing is just insane. If I lose my climbing partner, I've lost everything, because this person is my only help and support. He's the only one who can belay me on the next pitch, or help me survive another night. We need our partners, even if it's just to share the fear. It's not just one guy and another guy; it's a unit, a fellowship of the rope. The team is the sum of its parts, and it's not divisible. There's a feeling of solidarity, of shared identity, and it is unconditional.

H: The actions dictate the morals?
M: Not the actions. Being in a life-threatening situation.

H: But the imperative to act dictates the moral. It is a moral issue if I will share my water.

M: But it isn't imposed; it isn't dictated by external rules. Our behavior is instinctive. There's no right or wrong. I don't need any external rules and regulations. These things have been happening since time immemorial. Only mass-participation mountain sports need rules. And the seven billion people in the world need them, too.

H: As a young climber, your most important mentor was a man by the name of Sepp Mayerl. In what way was he better than others?

M: Sepp was several years older than me and could do everything. He was quite a naive sort of guy from a farming background, who worked as a roofer. He did churches mainly, all over the Tyrol. Without scaffolding. He used to drop a rope down these high church towers and repair them.

Anyway, this Sepp Mayerl was working in South Tyrol, and while he was there he wanted to do the north face of the Furchetta, a route I'd already climbed. But the day he went to do it, the weather wasn't good. On his way home he spotted my sister on the street and asked her if there were any climbers in the village. "Why?" "Well, I'm busy repairing the church tower, and I'd like to go climbing next week." And Waltraud said, "My brothers climb. They are extreme climbers." That's how we got together. He was seven years older than I.

H: What was the first route you did with the legendary Sepp Mayerl?

M: We went to do a route in the Sella region, an overhanging face, maybe 300 meters high. It was pretty amazing. I noticed straightaway that he could do a lot of things better than me. Sepp was also a really nice guy. He picked me up, drove me to the crag, and paid for my lunch. On the next route he let me lead a couple of pitches. "Go on, you have a try at leading," he said. On the third route we shared the lead—one pitch for him, one for me, leading through.

Sepp became my mentor, but he wasn't as fit as I was. That wasn't an issue when we were climbing; it just meant he couldn't walk uphill as fast as me. In short, we forged a partnership, and he became part of the team. Climbing in pairs, and often as a rope of four, we spent a few years ticking off all of the big challenges in the Dolomites. We did a lot of first ascents,

and second ascents of famous routes that had remained unrepeated for a
decade. We did some winter routes and went over to the Western Alps.
Those were good times.

H: Did you want to do routes faster than other climbers?
M: I followed human nature. I wanted to be closer to nature than the others.

H: Time wasn't a factor for you?
M: I knew that if I was fast, I wouldn't have to bivy on the route. That can
always be dangerous.

**H: Nevertheless, you were incredibly ambitious. Was it important for you to put
up new routes that were so hard that possibly no one else could repeat them?**
M: The desire to do that did surface over the next few years, yes.

**H: The major routes had already been climbed by previous generations. Was it
now all about finding harder new lines to do?**
M: That's right. Putting up harder routes on the famous alpine faces was
certainly an incentive. After a few years Sepp Mayerl said, "Our Reinhold
is taking too many risks." I used to solo extreme routes pretty often. We
were Sepp's favorite climbing partners, Günther and me—a couple of crazy
country boys from a small farm in the Villnöss Valley. Nothing was safe
from us: first ascents, second ascents, or third ascents. Back then, we were
regarded as the most creative climbers in the Dolomites.

H: You wanted to stand out from the rest?
M: Not really, but at the same time my brother and I had the idea of taking
things one step further. We started, quite deliberately, to avoid using bolts,
and to place as few pegs as possible. We were reinventing free climbing.
Our ambition was to put up new routes that couldn't be repeated, on big
faces—1000 meters or more—and with long run-outs between the protection pegs.

H: Why were you against bolts?
M: What's the point of them? I can't do the next moves, so I drill a hole in
the rock—a hand drill back then, a cordless power drill nowadays—and

hammer a bolt in. Anyone can do that. It means that any rock face can be climbed. Theoretically. But it's boring. And using bolts for aid means that nothing is impossible anymore. Which means the end of climbing development.

H: You hated all that ironmongery then?

M: At that time, in the mid-1960s, the development of aid climbing had reached its zenith. Back then, they were climbing featureless, overhanging 500-meter faces just on bolts. The first ones were placed 2 meters off the ground. The idea was, you drilled a hole in the rock, placed a bolt, and clipped an etrier—a short rope ladder—to it. And that's how you got up the route, from one bolt to the next. The drilling often took weeks of work.

You didn't climb the rock; you just climbed up the rungs of the etriers. You touched the rock, yes, but the natural features, the natural structure no longer played a part. What was important was the artificial aid, the artificial structures.

H: A kind of mountain autobahn. When did you decide it was boring?

M: I'd say 1966 or '67.

H: Apart from the boredom, were there any intrinsic reasons why you hated bolted aid climbing?

M: Well, you couldn't place any more bolts than one hundred in 100 meters. And the rock wasn't going to get any more overhanging. So I came along and said I'd climb without all that stuff. Progression, further development, was only possible if I renounced bolting.

In 1966 I wrote a little piece, an article that was published worldwide: "The Murder of the Impossible." I more or less said that if I use bolts, there's no such thing as "impossible" anymore, and without this, adventures are unthinkable. I only experience real adventure when I don't know what the outcome will be. Adventure means stepping into the unknown, or maybe the impossible. It's like being on another planet. If I do everything right I come back safe, if not then maybe I won't.

My radical stance set all of the "ironmongers" against me. Many of them still hate me for it, even today. There are two distinct schools of thought.

H: Your school wanted the direct duel with nature?

M: No, not like a duel. I expose myself to nature; I don't set myself against it or anyone else. I am prepared to step out of the bourgeois world and into a nonhuman world. When I am in that world, I can sense that humans do not really belong there. The others can keep on doing whatever they want.

H: Did you see your climbing as a form of art?

M: No, I never saw myself as a climbing artist. But I could spot a good line on the cliff that might offer a good free-climbing possibility. I was always looking for good lines. The line was very important. It was like an expression of my ability.

H: What is a good line?

M: One that adapts itself to the natural features of the mountain. Not the other way around, which means I don't adapt the mountain to suit my lack of ability. A climber who uses the natural features of the rock—edges, cracks, sequences of holds—is following a good line.

H: What is a bad line?

M: When I place bolts because I can't do the moves. I don't include the belay bolts on modern routes.

H: Because of your antibolt rebellion, some people derided you as a Neanderthal, who climbed mountains like an ape. How did you feel about this condemnation?

M: Apelike behavior patterns—it was an honor for us to be insulted like that.

H: It was said that you played a game of roulette when climbing, that you neglected to take safety measures on the mountain, that you were reckless, narcissistic, and had a death wish.

M: There was stiff competition between extreme climbers back then. But there were no serious disputes. What was discussed by people in the bourgeois world was irrelevant to the extreme climbers. We didn't take it seriously. They wanted security; I insisted on safety.

When someone has a bad fall, the good citizens will always call us idiots. For them the name of the game is: see the church from the outside, the pub from the inside, and the mountains from below. So it's your own fault;

if you hadn't climbed up there, you wouldn't have fallen off. My reply was, "You're right, of course, but beyond that, I'm not really interested in what you have to say."

H: For the extreme mountaineer, marginalization can be an accolade, a form of recognition?
M: We were way ahead of everyone else at the time. The recognition only came ten years later when clean climbing became the established ethic in the United States and young climbers couldn't get up our routes. "That's impossible, damn it!" the young stars said. "Either they cheated or they were mad."

In actual fact we were neither brilliant nor mad; we were maybe a touch better at free climbing than the aid merchants, and we trained a bit better. And we had this playful "let's just give it a try" attitude.

We tried a lot of things back then, but what is often forgotten is that we failed on a lot of things as well. Week after week we'd be out trying new stuff, getting so far and then getting stuck. I'd say that in 60 percent of cases, we'd manage to get up the route; the other 40 percent we'd have to retreat. Or not even start up the route in the first place. There were a lot of failed attempts!

H: How did you know when there was no point in continuing?
M: When we weren't in good form or when we got scared. Or because the weather wasn't good. Or because of something else. We often backed off things. Later, in my eight-thousander phase, about half of my attempts ended in failure and half were successful.

H: How did you know when it was time to turn back?
M: It was pretty quick generally. Either at the start of the route—it's too cold, I don't feel up for it today, the first pitch didn't go so well. Or maybe your partner would feel it was too risky. Günther often used to say, "Don't run it out. Put a peg in." So I would place a protection peg, try it again, get a bit higher but not much farther than that and maybe start thinking, "Will that peg hold a fall or rip out?" Sooner or later, you realize that today is not your day.

We were playing—but not in the childish sense. It was a game of skill. Will it work or won't it? That was the question.

H: You often slept badly before your climbs, or not at all, and dreamed about falling off.

M: Yes, that's the way it was for me. Maybe for others as well. I could talk a lot about nightmares, but only about my own.

H: That surely means you've been operating on only half power. Does it impair your climbing when you haven't slept well?

M: It can, yes, but I've noticed that if I don't worry about the route beforehand I don't climb as well. I think the adrenaline levels start to rise during the preparation phase.

H: Does the mental state that psychologists call "flow" give you additional energy when you are on the route?

M: Free climbing can create a flow state, but usually when I'm climbing below my maximum performance level and when I have the situation under complete control.

I've never climbed as smoothly when I've worked as a guide. Watching the clients often made me uneasy. They would grab holds that were loose. I'd be watching from 30 meters away, and I knew for sure that if they pulled on that hold, they'd pull it off. So I'd shout down to them to be careful. I wasn't a good guide; I prefer to climb on my own. When I'm climbing for myself, I ignore the loose holds, don't even touch them. I automatically use the ones that are solid.

Guiding clients also disturbed the rhythm of my own climbing, so when I went out with my usual partner on the weekend, it would take awhile to get back into that flow state again. The melody would all be wrong at first, and there would still be the remnants of bad climbing in me from all the guiding. Climbing is like ballet. Every second of the performance is different, as the structure of the rock determines how I compose and choreograph the moves. When I'm climbing well, I'm not thinking, not at all. It's all instinctive, and if I find the right flow, the climbing just happens of its own accord. It's as if gravity has been abolished.

H: You spent your adolescence climbing rocks. Didn't you miss the girls?

M: The climbing was more important. I guess I vented my sexuality on the rocks. My fantasies were stimulated much more by new routes than by sexual conquests and "belay chicks."

H: For many twelve- or thirteen-year-olds, that is the number one topic.
M: I guess I was a late bloomer. We war babies achieved sexual maturity much later than kids do these days.

H: Were there any girls in your valley that you found attractive?
M: In my valley, at that time, no. At the end of the '60s, yes, but I'd already left the valley by then. I was in Bozen, Eppan, Padua. There were girls at college that I admired. From afar.

H: Were you shy?
M: Very shy.

H: You didn't know how to talk to girls?
M: No.

H: Why didn't you try to impress the girls with your climbing skills?
M: The way I saw it, there was the climbing world, which I shared with a few close friends and my brother, and the bourgeois world. I didn't want to share the former with school friends, a girl, or anyone else. There was this feeling that climbing had nothing to do with them. They had no idea what it was all about. Same as my climbing partners, who were all older than me—they never asked me about my school friends or what I was doing at school. Again, that was nothing to do with them.

H: Were you a kind of sect?
M: We were a sect, yes.

H: And you didn't want the sect to become defiled or adulterated?
M: We were a small circle that was not defined by written rules or statutes. By mutual experiences maybe, shared stories. And by what we did. We had no rules, only similar experiences and shared adventures, and anyone who hadn't had these experiences couldn't be part of it.

The schmoozers and bullshitters, who were so full of themselves—and there were enough of them—we treated with contempt. For us, what you did was what counted. If you'd done something, you had something to say. One of us only had to mention the Eigerwand, the third pitch on the ramp, the overhang on the left, and everyone knew what he was talking about.

H: You had to have frozen together and suffered together to belong?
M: Yes, because no one can imagine what it means to freeze through a night on the mountain at minus thirty degrees if he's never experienced it. Storms, losing your way, death—these are all things I could never explain to the uninitiated.

What also helped to bind this elite group together was a feeling of self-assurance. There was no arrogance, but the feeling was that if the world ever got difficult to deal with, we were the ones who would manage to cope. One reason being that other people had never learned how to survive in a difficult world, whereas we had survived in situations where survival seemed impossible. That's where the self assurance came from. In normal daily life, too. It wasn't about superiority—you weren't any cleverer, any better, or any richer. But you were better able to survive.

It wasn't a cultural thing either. On the contrary, we went to the theater less often, read less, and had less money than our peers. We'd learned how to survive, and it was this that gave us the self-assurance that made us strong in normal life. I was never scared that I might not get my life sorted out, whereas my father was always scared. He was even scared on my behalf—that I'd end up on the streets and so on. Just because I hadn't been studying and had spent the whole summer climbing.

H: Do you remember when you first kissed a girl?
M: There was no big love before 1971. I slept with a few women, of course, but there wasn't one definitive love story.

H: How would you describe it then?
M: None of those relationships was important enough for me to give up my fanaticism for climbing.

H: Was your time too precious to waste?

M: Yes. It would have meant losing a weekend. I get bored just hanging around.

H: Going for an ice cream with a girl, for example?

M: That would have been unbearable.

H: What was so bad about it?

M: Nothing. But that kind of thing has always been a problem for me. Having a chat, sharing a bottle of wine, or even two, climbing a mountain together, or looking at a painting, yes—but not going for an ice cream. Killing time gives me the horrors.

H: Having an ice cream and watching the world go by?

M: Absolutely unbearable. I do things with a passion or not at all. Going for an ice cream with a girl meant nothing to me, even as a young man.

CHAPTER II
HIGH-ALTITUDE
MOUNTAINEERING

1969–1986

*An irresistible urge drives me on,
compelling me to attempt higher and harder things,
to push myself to the limit.*
—Hermann Buhl

STORM AND ICE

I got into high-altitude mountaineering by chance. In 1969 I was in college when I joined a Tyrolean expedition to the Andes because Kurt "Gagga" Schoisswohl, an excellent rock climber, had dropped out. It was a good year later that I returned to the University of Padua, after a tragic expedition to Nanga Parbat and the amputation of several toes and fingertips, and I found myself unable to continue my studies. So I became a high-altitude mountaineer.

My rock climbing was hampered by the amputations, and in particular by the pain in the ring finger of my right hand, so from then on I concentrated on the big mountains. I traveled a lot, worked as a group leader to earn a living and to finance my own trips, and wrote books and gave lectures about my expeditions. And soon I became the representative for a generation of climbers and travelers with a thirst for adventure but without the means or the skills required to climb these high peaks themselves, which were accessible only to the few. I wanted to gain experience and to report on my trips. For me, it was always about that state of uncertainty that exists between life and death, and the helplessness of human existence at the top of the world.

Nowadays, adventure is often confused with getting quick-fix kicks. Rock climbing and high-altitude mountaineering have become tourist activities, and even Mount Everest has become a commodity, bookable for a long holiday, an all-inclusive trip with entertainment, insurance, and oxygen depots on the summit. I am not complaining, but one thing is certain: without self-reliance and exposure to risk, the experience of climbing at altitude is very different; you might as well be in a kindergarten class.

H: In 1969 you left the Alps for the first time and went on an expedition to the Andes. In what way was that trip a career-changing event?
M: It turned out to be quite a spectacular change, but that aspect of it wasn't important for me at the time. I never really thought about careers.

What's more important is the fact that at that moment I finally gave up the idea of leading a middle-class way of life. I was totally unhappy at college. I somehow had the feeling I was missing out on life. I had been trying to finish my engineering course with the best intentions in the world, but I was just forcing myself to do something I didn't really want to do.

And then these Innsbruck climbers called me, three days before they were due to leave for South America. They had been planning their expedition for a long time, but I didn't know much about it, only that my climbing partner Sepp Mayerl was on it, and Peter Habeler, the brilliant Zillertal mountain guide. I knew both of them well from the routes we'd done together in the Alps. I was told that someone had dropped out, so I could come along if I wanted. Tickets, equipment, everything was sorted out, and the clothing would fit me more or less. I wouldn't have to pay for anything. I had three days to get a visa.

The Andes expedition was the perfect thing for me. I came home fit and more experienced, having climbed two big mountains—Yerupaja Grande and Yerupaja Chico—by new routes, and I was hungry for the Alps. I subsequently set about breaking some of the last big taboos in alpine mountaineering.

At the time, 1969, the north face of the Droites was the hardest ice climb in the Alps. It had only been climbed three times, and never without falls. The first ascent took six days; the fastest, three days. I had attempted the face with my brother in 1965. We got scared and backed off it. Since then, no one had been able to climb it.

H: Pure ice?
M: Ice lower down, mixed climbing—rock and ice—above.

Back then, we just had ice axes, no modern ice tools. I set off early, at first light, with my axes and crampons, and a rope tied around my waist. Nowadays the route is no big deal, but back then it was scary. By midday I was at the top, watched by some aspiring mountain guides from Chamonix. That got me known in France and brought me my first advertising contract. It was the start of my professional climbing career.

H: Who did you thank when you got to the top? Sepp Mayerl? God?
M: Neither of them. I did feel a kind of gratitude, for my good fortune and

the courage to do what I did. Then it was all over, and I just cried with joy. But there was no desire to head out the next day and push things another step further. It was okay as it was. There was no God involved, so why should I thank him? And I'd certainly pushed things much too far for my mentor to approve.

H: But you were hooked?
M: Hooked on what? I only wanted one thing: not to die.

H: You were addicted, a climbing junkie. You'd been high, you'd had your fix, you were happy, you were content, you were calm.
M: Yes, I was totally satisfied—calm and excited at the same time.

H: You went back down and thought, "I won't be doing anything like that again in a hurry."
M: I went down and thought, "No one is going to repeat that anytime soon, not even as a rope of two."

H: And two days later you were off again.
M: A day later. The Freney Pillar.

H: How did Günther cope with all this?
M: He was a bit cross in the summer of '69, because I didn't climb very much with him and all the soloing I was doing bothered him. He said, "When you ask me to come climbing with you, I can never go. If only I didn't have to work in that bank!"

H: You were spending more and more time in the mountains, whereas he could not decide between the mountains and his job at the bank.
M: A necessity.

H: He was torn between the two.
M: Yes. He then decided to quit his job for the Nanga Parbat expedition. He wasn't on the team at first. He joined us because others—Mayerl, Habeler—dropped out.

H: That was the beginning of the catastrophe.

M: The Nanga Parbat trip was the watershed, a demarcation line between our youthful exploits as idealistic, obsessive young rock climbers and high-altitude Himalayan mountaineering. I would never have gone to Nanga Parbat just to be on the third expedition to reach the summit. The Queen of England herself could have invited me, but I wouldn't have gone, not just for the summit. But the Rupal Face, the hardest face on Nanga Parbat and the biggest face in the world—now that was a real challenge. It was like three Eigerwands stacked on top of each other, at altitude—the next dimension of mountaineering.

In the Alps I had done all the things you weren't supposed to do: the hardest ice route, solo, without a rope; and the hardest rock route, solo, with back rope protection on only a few sections. Both were big, dangerous 1000-meter routes. I just went and did them. Around that time, the mountain journalist Toni Hiebeler, a clever writer and chronicler, wrote me a letter: "If you don't stop, you'll be a dead man by autumn." I knew that no one else had the nerve to do what I was doing. I was so self-confident. Not better than the rest, but instinctively good. I knew that at that time no one could do what I was doing. When everything flows, it's like a dream. Everything is so natural.

That autumn, the Nanga Parbat invitation arrived. I knew, of course, that the best German climbers had failed on the Rupal Face in the 1960s. Nanga Parbat wasn't just the "Mountain of Destiny" for the Germans. It wasn't just the mountain that Hermann Buhl had made the first ascent of—a brilliant achievement. And it wasn't just one of the most difficult 8000-meter peaks. Nanga Parbat was where *the* biggest face was to be found: the unclimbed Rupal Face. It was 4500 meters high and vertical at the top—the face that Hermann Buhl had said was impossible and would never be climbed.

H: Hermann Buhl was your great role model. His assessment didn't scare you off?
M: Buhl went on to say, "Even to attempt it would be suicidal." I viewed that as a challenge. Seventeen years had passed since the first ascent of the mountain, and a new age had arrived. We had better equipment, and we knew more. For me, it was all about that face, and that face alone.

H: You received the invitation from a man by the name of [Karl] Herrligkoffer, a strict expedition leader with Greater Germany tendencies, a man whose half

brother, Willy Merkl, had lost his life on Nanga Parbat in the 1930s. Were you not a little uneasy about going with Herrligkoffer?

M: I had every reason to be, and my decision to go was stupid. After the 1953 success, Herrligkoffer really slaughtered Buhl, his star climber. He was also driven by the spirit of the 1930s expeditions, and had sworn "never to rest until the swastika flag flies on the summit of Nanga Parbat." But the worst thing was, he had no idea about climbing big mountains. He wasn't a very good organizer either. And he retained the sole rights to publish anything about the expedition.

Naturally, I had no idea that things would turn out as badly as they did. We just pushed all the hassles to the back of our minds, took a deep breath, and got on with it. We wanted to go to Nanga Parbat. My brother Günther was right when he said, "We would have gone to Nanga Parbat with the Devil himself."

H: But Günther was not on the team to start with.

M: My father really wanted me to take him with me, so I made an effort to get him on the team.

H: Nevertheless, Herrligkoffer had not bargained on taking Günther, so you must have feared there might be trouble in store.

M: Herrligkoffer always talked about comradeship, but he himself wasn't comradely in the sense of sharing, fellowship, and identification with the common objective. Nanga Parbat belonged to him and him alone.

It started with us having to stay behind him on the walk-in to base camp. It was a real ordeal, as the man needed an hour to walk the same distance I could cover in ten minutes. I couldn't stand it, so I went on ahead and kept getting told off. Single file rather than on your own—those were Herrligkoffer's orders.

H: Before the climb you sent a postcard home. You wrote that high-altitude mountaineering was a real grind.

M: It is, too. Your throat is generally swollen and phlegmy, and you have a cough because you're gasping for breath and the air is so dry. Your food tastes like nothing. It's too hot during the day and too cold at night. The radiant heat saps your strength; you get a bit apathetic. High up it's bright, but the sky is black. And when you look down, most of the time all you can see is a kind of gloom, rather than the valley bottom.

That face is an abyss; it's impossible to tell whether it's 2000, 3000, or 4000 meters from bottom to top. There's just a gaping void all around you. But I wasn't scared. I'd done 1000-meter overhanging walls in the Alps and felt no anxiety.

H: In Herrligkoffer's assault plan, you were the lead climber for the summit push, while your brother Günther was to fix ropes at 7500 meters for days on end to safeguard the descent. Did your brother feel cheated of the summit?
M: Günther was annoyed. He didn't think he would be given a chance to go for the summit.

H: Did he feel like Herrligkoffer's slave?
M: Günther was clever enough to know that plans can change right up to the very last moment. No one could know at the start of the expedition who would drop out on the way to the top. As it happened, nearly all of them dropped out. It wasn't as if Günther and I simply grabbed the summit for ourselves.

H: But [Peter] Scholz and [Felix] Kuen were up there, too.
M: In our tracks. Two strong alpinists with backup and plenty of courage. Nevertheless, if we hadn't found the route and broken trail, they wouldn't have stood a chance. They probably wouldn't even have tried.

H: You were up at the top camp with your brother and the mountain film-maker Gerhard Baur. What was it that made you decide to go for the summit on your own?
M: The weather was threatening to turn bad. New snow and avalanches would have made climbing the Merkl Couloir impossible, so I suggested to Herrligkoffer that I set off at night and get as high as I could during the following day—maybe even all the way to the summit—before coming back to the camp, where my brother and Baur should wait for me. "My sentiments exactly," was Herrligkoffer's reply.

H: The agreement with Herrligkoffer was that they would fire a red rocket from base camp if the weather forecast was bad, and a blue one if the forecast was good. Red meant you could go on your own; blue meant you should only attempt the summit as a team. The weather forecast was good, yet Herrligkoffer fired a red rocket. Why?

M: I don't know. They still say it was an incorrectly labeled rocket. Maybe it was.

H: Your expedition companion Max von Kienlin maintains that you knew the weather would stay fine.

M: And how was I supposed to know that, may I ask? I had no access to the weather report. We didn't have a radio up there. That's why we agreed on the rocket signal. Von Kienlin makes a lot of things up; he lies.

H: What did you see when you set off at three o'clock in the morning?

M: A vertical wall above me, and a clear, starry sky. It was icy cold.

H: Good weather then?

M: Yes. Otherwise, I wouldn't have set off. But that meant good weather only for the time being. It started getting foggy again in the afternoon, and the next day, on the descent from the summit, it was even snowing below us. The poor weather forecast was the decisive factor for the decisions I made in the summit region.

H: When you left Baur, you say you told him, "Wait here for me until I get back." But Baur says he heard nothing, that he was asleep.

M: No one sleeps up there. You doze. I gave my instructions the evening before I set off: "Fix ropes and wait for me!" Baur confirms that I asked them to wait for me, and has done so for thirty-four years. I asked them to wait for me during the evening of June 26.

And what about my second request? Why would I have asked them to fix ropes in the Merkl Couloir if I hadn't intended to come back down that way? Why should I have lied to my brother and Baur?

H: What did you take with you for your solo push?
M: Several layers of clothing, spare gloves, and a small tube of vitamin tablets. That's all.

H: How did it go?
M: It went well, but I kept having to traverse to avoid the steep sections. The routefinding cost me valuable time. And concentration. I saved the others a lot of hard work. They didn't have to scramble around on confusing terrain searching for a route.

The route got easier toward the top, but it was still clear to me that the couloir would be pure hell if it snowed—no chance of survival. There would be snow coming down it continuously, a raging torrent that would sweep you off the face.

H: Your brother was supposed to wait with Baur at the high camp. When did you notice that he was following you up?
M: I don't remember the details, but that isn't important. What I do remember is a rock spike over to the left—I can still see it in front of me—and then he appeared.

At first, I didn't know it was him, but then I recognized him by the way he moved. So I waited. On the huge ramp that leads diagonally up to the south shoulder.

H: What did you say to Günther when he joined you, contrary to your agreement? Were you angry with him?
M: No, I wasn't angry, just irritated. For a moment I was shocked, because his appearance broke my concentration.

My objective was to make it to the summit and back in that one day. Climbing as a pair is different; you are constantly keeping an eye on your partner. So my first questions were "Did you bring a rope?" and "Did you fix the couloir?" When Günther said no, I knew the whole thing might get critical—far more dangerous than climbing solo.

The previous summer I had climbed some of the most difficult faces in the world on my own. Günther had problems with that. He'd told me a few times, "I don't like it. I get scared when you do your crazy solos."

H: Why didn't you send Günther back down?

M: That was out of the question. We had been climbing partners our whole lives. On this expedition Herrligkoffer had blocked Günther's chance of a summit bid. Now my brother was seizing that chance for himself. He had also proved he was capable of doing it. He had climbed the Merkl Couloir faster than me. Günther had positioned himself for a summit attempt.

I could have given him an ultimatum: "Either you go back down or we both go down." But I didn't do that. We both saw a chance of getting to the summit.

H: But in spite of this, you still had a problem with Günther being there?

M: For sure. If I'm on my own and I die, then I die. But I couldn't allow Günther to die; I had to get him back down safe and sound. I was older, so I was responsible for him. It wasn't the same as it used to be; it wasn't as safe. I was more experienced, a soloist, the older brother, the one who had always done the leading.

So yes, I was my brother's guardian; of course I was. That's why it is so inhuman when certain expedition members say, "He didn't look after his brother. He simply sent him back down the Rupal Face." I didn't send him back down, either before or after the summit.

H: When did you notice that Günther was tired?

M: On the descent. He kept stopping to take photos. It was an excuse to rest. I know the score. If someone is having trouble keeping up, he never comes straight out with it and says, "I can't go on."

All this happened on the snow ridge on the way down from the main summit to the south summit—an easy-angled ridge, firm snow, a straightforward walk. Even on the ascent—I was excited, as I could see the Silver Saddle for the first time, below me, and big banks of cloud—Günther was getting slower and slower. It took us an hour to do the last 50 meters. On the top I had the feeling we were both equally tired, exhausted by the climb and especially by the sun. I kind of realized we wouldn't make it back down that day. But I thought it would still be possible to get down some of the way at least.

H: What image do you still have in your mind of the summit?

M: My brother and the Silver Saddle, no more than that. Plus a feeling of absolute responsibility for my brother.

H: How did the first part of the descent go after the summit?

M: Günther was lagging behind more and more. On that easy snow ridge. When we got to the south summit, where the face drops away almost vertically, he said, "I'm not going down there. I can't do it." That's when I knew it was going to be difficult.

I couldn't tell him he had to do it, knowing that if he fell, he'd fall to his death. You don't say that to your brother when he's exhausted and dazed.

H: How long did you spend on the summit?

M: An hour perhaps, then we started down.

H: Was it still light?

M: Yes, it was still daylight.

H: How long does it stay light on Nanga Parbat in early summer?

M: We could see, very well in fact, until we got to the bivouac at the Merkl Gap. We would certainly have gotten about halfway down the Merkl Couloir in daylight if we'd chosen that descent route, but we chose differently. We decided against the south side because Günther was scared of that descent, because we were expecting bad weather, and because the face drops away vertically there.

I'd noticed that Günther was not moving as safely; there was no 100 percent guarantee that he'd be able to downclimb something as steep as that. He was staggering and kept squatting down; these were bad signs. He was slow, too, much slower than he'd been on the way up to the summit. And the hour's rest on the top hadn't helped him regain his strength; on the contrary, it had weakened him even more. It was a combination of the altitude and tiredness, of course, after the long slog up to the last camp the previous evening. We'd been climbing for a day and a half with nothing to eat or drink.

H: But you had some dried fruit and vitamin tablets, didn't you?

M: I had vitamin tablets with me, effervescent tablets that you dissolve in water. But we didn't have any water. Later, on the descent, I tried to dissolve them in melted snow, but it didn't work. It took far too long, and the foam was undrinkable anyway, just a thick broth. It was too cold, so you couldn't melt snow in the tube in your hand. If only we'd been able to get a mouthful of soup down us.

H: And the dried fruit, how was that?
M: Emergency food, nuts and raisins, but not even a handful each.

H: How did you come to the decision to descend a face that had no fixed ropes and no camps on it? It was totally unexplored territory for you, it was known to be avalanche prone, and there were no expedition members camped on the face who might have been able to assist you at some stage.
M: The terrain on that side really was an unknown quantity, but we could see that it wasn't as steep as the way we'd climbed up. We also knew that the section down to the last camp on the Rupal Face hadn't been fixed either, and that section is really steep and difficult, plumb vertical in places. One slip there, one false step, and you'd be off.

H: You didn't have a rope with you?
M: Neither of us had a rope. That's why my brother was so unsure of himself. He said he didn't dare to climb down that face without a rope. We didn't talk much. But that statement of his was definitive. He wasn't going to descend that way; it was too dangerous for him. We both knew there were no fixed ropes above that top camp. There was nothing to discuss.

Of course, there were other expedition members, and tents, on the south side, and if we could have gotten down to 7300 meters, Günther would have been saved. But first I would have had to get him down, and that was still a height difference of 700 meters. Without a rope it was unthinkable. It was also clear that we wouldn't make it down that day. It was too late for that. That was our own fault, of course, but the fact remains that it was too late.

So that meant we only had two options: either go back down the steep south side and run the risk of one of us slipping off, or descend a little way down the easier angled northwest side. I wasn't too worried about the steeper route. I told myself we'd probably manage it somehow. But my brother refused.

H: What did he say?

M: He said, "I'm not going down there. It's too dangerous, too scary."

H: For years you had always been the driving force in the climbing partnership, yet in this situation you left the decision to your brother?

M: On the contrary. But I had to take his fears seriously. We both considered what to do, inasmuch as we could think clearly at all, that is.

H: Why didn't you think, "The weather is good, our companions will come soon, and that might mean we can get a rope and get down tomorrow"?

M: That scenario never crossed our minds. The information we had was: bad weather coming. Just that.

H: Nevertheless, didn't you have any confidence that other members of the team would come up?

M: The idea that someone would come up was out of the question, unthinkable. After the red rocket, I'd set off for one final attempt. Climbing the couloir in a fall of snow would be tantamount to suicide. Why would anyone do that? The possibility of anyone climbing a little way up the route from the top camp could be ruled out, for the time being at least. The following day it was conceivable. The next morning, June 28, I hoped that someone might do just that. Surely they must be missing us at that point? And the weather seemed to be holding, too, so far at least. But we now had no chance of retracing our route.

Why, you might ask, didn't we bivy on the summit? That, too, was out of the question. It was the worst possible place. The summit ridge is very exposed: there is no shelter; it's 8000 meters up and exposed to the wind. It's extremely cold that high up. I had a photo of the Rupal Face with me, logically enough, and got it out to look at the summit region. Where could we descend? What should we do? Those were the questions. I compared the photo to the reality. That's something climbers often do. Every experienced climber studies routes on photographs.

H: Why, if it was never your intention to traverse the mountain, did you take a photo with you?

M: I still have that Rupal Face photograph. It was the title shot of a Karl

Maria Herrligkoffer book. About as big as the palm of your hand. It shows the whole of the Rupal Face up to the summit, 5000 meters of face in 180 square centimeters.

My instinct and my routefinding experience told me it must be possible to traverse from the Merkl Gap into the Merkl Couloir. Only then did I consider descending as far as the gap. I also thought that, being lower down, we'd have more shelter there, and more oxygen. It wouldn't be so cold down there either, and it would be less exposed to the wind, so we could bivy there. When I looked toward the northwest, I immediately saw that the terrain wasn't as steep. The descent to the Merkl Gap had to be possible, I thought. So we gave it a try. We headed down, step-by-step, over relatively easy ground.

H: What grade was it?

M: Hardly worth mentioning. Grade I, maybe. We were good climbers; it wasn't a problem.

If we really had intended to descend the west side at that point in time, then going down to the Merkl Gap would have been stupid. The logical thing would have been to head down directly to the west. The simple, easy-angled way down, hands-in-pockets terrain, lay directly to the west. But we only descended to the west for a short distance and then turned left, onto steeper ground, in order to reach the Merkl Gap, the intention being to find a way back onto the Rupal Face the next day.

It became clear that getting down any farther than the gap that day was out of the question. We got to the gap around seven or eight. We didn't bivouac at the gap itself but a little way before it, where there was more shelter. We wanted to get back onto the Rupal Face the following day.

H: There are those climbers who maintain that the upper part of the Rupal Face is not particularly demanding and that the difficulties your brother was worried about actually start much lower down—below the point at which you intended to rejoin the Merkl Couloir.

M: Firstly, our descent route was several grades easier than our route of ascent, even on the upper part. And secondly, we had to get out of the death zone as fast as possible; in other words, we had to lose height fast. Then there was our agreement with the weather and the rocket—the red rocket meant we would have had to downclimb the whole of the summit wall with no ropes and no protection.

H: Nevertheless, it was still far from certain how you would get back into the Merkl Couloir.
M: For sure. But first we had to bivouac at the gap. I was sure we'd find a way back into the couloir. That's why I didn't go all the way to the gap that day to look down the Rupal Face. It was only early next morning that I saw it wasn't possible to descend from there. You couldn't downclimb from there.

H: Why not?
M: From above, it all looked vertical, covered in windblown snow.

H: Vertical ice and vertical rock?
M: Vertical rock with a dusting of fresh snow, windblown snow. Below me there was nothing but fresh air, a bottomless pit, with a few dark patches—jagged rocks—poking out here and there. Far below I could see the Merkl Couloir, which gets less steep toward its end, maybe sixty degrees. The tracks we'd made the previous day were still clearly visible. It was immediately obvious to me that there was no point even trying to downclimb from the gap to join those tracks. Without a rope it was unthinkable.

H: How far down was it?
M: About 80 to 100 meters. And vertical, almost all the way down to the section we'd climbed past the previous day.

H: Absolutely vertical?
M: It was vertical, yes. It was like I was standing on a high-rise building, on the edge of the roof, looking down. Any normal person would have felt nauseous looking down there, scared of losing their balance. I kept leaning forward and peering down. I could see it was impossible to climb down there. I didn't need to ask my brother if he dared to give it a try. I didn't trust myself to do it either.

H: You must have despaired when you saw what was down there.

M: It was a shock, yes. I knew then that we were on our own, at the mercy of the mountain. The question was: What could we do about it?

H: You were misled by the photo and made the wrong decision?

M: Yes. Although the photo did suggest a possible descent route, a traverse across a snow-covered ledge system. Either the photo was taken after the wind had plastered that steep wall with snow or there were huge cornices on the ridge that didn't throw any shadows. Anyway, it was too late for questions and speculation.

H: When you realized you had made a mistake, you had already spent one awful night on the mountain. Günther was delirious—something about a blanket?

M: He kept going on about a blanket that wasn't there and reaching out for it.

H: All you had were lightweight space blankets, which you wrapped around your feet, and you took off your outer boots and sat on them. Can you describe the bivouac site?

M: It was a flat place among the rocks and snow, a hollow below the summit wall. We sat on our outer boots for a bit of extra insulation.

H: Did you have your backs to the rock?

M: Yes. On one side there were a few boulders; the other side was open to the elements. But the wind still got to us. There's no avoiding it up there. At least we weren't crouching on an exposed ridge, though, or in the gap itself, which would have been really drafty. It was about the best spot we could hope for in the immediate vicinity.

H: How cold was it?

M: We didn't measure it. If I say it was minus forty, that might be an exaggeration; if I say it was minus thirty, that's probably an understatement. It certainly couldn't have been much colder.

H: Thirty degrees below feels even colder at altitude.

M: That's because your blood supply is less efficient. Even if you have enough blood sugar, your body can't produce any heat because of the lack of oxygen.

At 7800 meters the cold feels much worse than it does at 2000 meters above sea level. Your body produces heat by burning sugars, but at altitude your metabolism slows down.

H: Did you tell Günther that there was no blanket?
M: I did, yes. We didn't talk much, though. Just a few words now and then.

H: Günther was hallucinating, and that was a real cause for concern. Could you feel yourself starting to panic?
M: Well, I knew you could get hallucinations when you were that high up, and I thought that must be it. For me, that was an indication that Günther was suffering from altitude sickness, although altitude sickness is a broad term. If I get a headache at altitude, I might have altitude sickness.

H: Nevertheless, you must have been getting more and more worried about Günther. He couldn't walk, and he was hallucinating.
M: First of all, I said I'd go and see if and where we could get down. Then I went back and told him it wasn't possible.

H: How did you fight down the rising panic?
M: The panic only came later. Neither of us was the type to give up right away. Okay, so we couldn't get back down, but we weren't about to just sit down and die.

Anyway, we still hoped that we might get help. We hadn't returned to the last camp as planned, so the others might have climbed up a way to look for us, to help us. Or maybe they were at the camp and, if the wind was favorable, they might be able to hear us. That's why I started shouting. Even though I knew full well when I looked down that steep wall that we couldn't climb down, I still hoped that someone might be able to climb up to us and bring us a rope so we could get back down to join our ascent route.

H: You wrote about the cold being so bad that you could move your toes but couldn't feel them. It was all just pain and fear up there. Yet you still spent three hours shouting for help.
M: Yes, but not continuously. I didn't leave my brother on his own for three hours. I went to the gap, shouted, then went back to the bivouac site.

H: And what was Günther doing during that time?

M: He just crouched there and asked if anyone was coming. Until someone did come. Around nine o'clock some people came into view.

H: Could you make out who it was?

M: Not right away, but then I recognized them.

H: How?

M: By the way they were moving and the color of their clothing. Kuen is big and very slim. It looked like Scholz was with him. But it was Kuen I recognized first, by his movements, his size, and his rucksack.

H: Were you pleased that rescue was at hand?

M: I wasn't euphoric, but I did tell myself that there was hope for us now. If we were lucky and the two of them managed to make it up to the gap, we wouldn't exactly be out of trouble but at least we'd be reunited as a team of four and we'd be able to use their ropes to get back down to the last camp.

H: It was then that you noticed that a storm was coming in?

M: The moment I saw that they were getting closer, I stopped shouting for help. The sky was clear, and even Günther thought we'd be rescued. Nobody carries on shouting for help once they can see that help is on its way.

H: But you still didn't know if they had seen you. After all, it was very windy and there was a lot of cloud.

M: We weren't stuck in the mist; we could see each other. There was a lot of waving and shouting.

H: So you waved and they waved back, but communication was difficult?

M: Yes, communication was difficult, of course. Anyway, as far as I was concerned, they'd come because of us, because they'd heard my shouts for help.

Felix Kuen even tried to climb up toward me. He disappeared from view and reappeared in the couloir. He couldn't manage it. He wrote in his diary that even to attempt it would have been suicidal. Kuen kept trying and going back down again, and it slowly dawned on me that he wasn't going to make it.

The weather wasn't too good now, but it looked like it might just hold for another few hours. The clouds were building—that was the bad weather that was forecast, I thought. Later that day—June 28—it snowed.

H: There then followed one of the most famous dialogues in the history of moun-taineering. Kuen asked, "Are you okay?" and you replied, "Yes, everything's okay." Had you taken leave of your senses?
M: Before I gave up, I had one last try at directing them up to the gap to join us. If they were intending to go to the summit, I thought, they should not feel obliged to take us down to Camp 5. All they had to do was come up to us and leave us a rope. We couldn't have climbed back up, but we could have rappeled down. Even half of the rope would have been enough for the couloir. From the gap, Kuen and Scholz would have been able to get to the summit relatively easily by following our descent route, which was much easier than the route we'd taken to the top the previous day. They just had to make it up to the gap first. Those 100 meters were the crux.

When I realized they weren't reacting to my signals, I told myself that the section in question must be impossible to climb. If they'd ignored common sense and tried to climb it and fallen off, Günther and I would have been dead anyway. The rescue plan was no good to anyone if it couldn't be done.

It was only then that I waved them away. "Everything's okay" meant "Go on; don't worry about us." It also meant I was removing the burden of responsibility from Kuen and Scholz, as if I'd never shouted for help in the first place. If I hadn't told them to go on, they might have tried to get down to us from the summit. They would have been twenty hours too late. If we'd stayed where we were, we'd already have been dead by then. And if we decided to go down a different way, they wouldn't be able to help us anyway. They might even have died. That's why I felt compelled to indicate that everything was okay after our shouted conversation.

H: It wouldn't have taken them twenty hours, would it? Kuen wrote later in his expedition report, "We would have helped. We would have taken about five hours, but we would have been there with full equipment."

M: How? They were later than we had been. If they had followed our route, they'd only have reached the south shoulder in the evening. Their variation was easier and shorter, yet they'd still have had to bivouac on the way down below the south shoulder.

H: In his diary, Kuen went on to say, "We wouldn't only have *been able* to help, we *would have* helped."
M: The only route I knew about was the one we'd taken. The route they eventually opted for turned out to be shorter and easier. I congratulate them on finding that alternative route, but I didn't know about it at the time.

If they had followed our route and climbed at the speed they were moving before our shouted conversation, they would have reached the south shoulder in the dark. If they had come down to us straightaway, either after the summit or directly from the south shoulder, bypassing the summit, they would no longer have been able to help us that day, June 28, even if we had waited for them. The four of us could have spent the night up there, but Günther wouldn't have survived that second night, and there might only have been two, maybe three, people making the descent the following day. Or maybe nobody. That would have been of no use to Günther.

H: Nevertheless, if you had shouted for help, Kuen might have appeared.
M: But not until that evening at the earliest. In other words, too late. The facts are clear. With hindsight, waiting at the Merkl Gap was our biggest mistake. But we were hoping that help would arrive, and we wanted to get back onto the Rupal Face.

H: Were you worried that you might not survive a second night that high up?
M: Another night up there would have been fatal, definitely for my brother and probably for me as well.

H: And a rope would not have helped?
M: Not any longer. Why should I risk my own life to lower a dead body down the mountain? I could have left the corpse up there.

H: You could have used a rope to rappel down into the Merkl Couloir. Maybe that same evening.

M: Nobody could rappel down there in the dark. It would be much too risky. If Kuen and Scholz had been able to get to us around midday, or early afternoon at the latest, we might have had a slim chance. But there was now zero hope of that happening.

Why should I have assumed that they would climb twice as fast after we had seen them than they had before our shouted exchange? We all get slower and slower as we approach the summit. Kuen and Scholz also got slower.

H: What was the last you saw of Kuen and Scholz?

M: I watched them as they kept going upward, then I walked away from the gap. The moment I said, "Everything's okay," I severed our emotional connection; it was like cutting a cord. I didn't even think about what to do next. Anyway, they couldn't help us at that point. They'd done all they could.

H: Why didn't you climb back up to meet them and get the rope?

M: Going back up was out of the question. We were too weak by then. Even I wouldn't have managed the 250 meters of height gain to the south shoulder. Or maybe I'd have gotten there in the evening.

And what about Günther? I would probably have died of exhaustion up there, and he would have frozen to death in the meantime. It was out of the question. The only thing that is relevant is what we actually decided to do.

H: What did you decide to do?

M: The only way out was to climb straight down by the easiest route possible.

H: It was unfamiliar territory. How did you find your way?

M: There were two imperatives: to find a descent route, and a route that we could downclimb given the state we were in. I decided to head for the Mummery Rib, because I knew that was the line chosen by [Albert] Mummery for his attempt in 1895. I didn't know exactly where his line went, or how far up it he had gotten, but I figured that whatever Mummery had climbed seventy-five years ago we should be able to manage without a rope—theoretically, at least.

H: There was a storm during your descent. How far apart were you and Günther?

M: A hundred, two hundred meters. Sometimes more. Three hundred maybe.

H: You were doing the routefinding?

M: Yes, I was ahead, looking for the best way down, and it was only when I was sure it was okay that I let Günther follow me. I directed him from below by shouting and waving. Sometimes we were quite a long way apart, because I wasn't always sure which was the best way to go.

H: How severe was the storm?

M: The storm was below us. Luckily, it didn't snow heavily; otherwise, it wouldn't have been possible to keep going down, as the risk of avalanche would have been too high. We had a really firm snow base, so the going was good for a while, but at the start of the Mummery Rib there was hardly any snow, just bare ice. If we wanted to avoid retracing our steps, which would have involved an interminable effort, the rib was our only option. So down we went, step-by-step.

H: You then had to do the one thing you'd wanted to avoid—bivouac for a second time, even though Günther said he couldn't take another bivouac.

M: The second bivouac was a totally different situation: much lower down, nowhere near as cold, on rock. It was obvious that a second bivouac was in the cards, as tired as we were.

If I'd been on my own, I'd have gotten down faster. Whether or not I'd have managed it in a day, I don't know. On my solo ascent of Nanga Parbat in 1978, eight years later, it took me six hours to get down, although I certainly wasn't in any better shape physically than I had been in 1970. That's six hours for a descent that took Günther and me a day and a half—three times as long.

H: How long did you bivouac?

M: From midnight to dawn. The problem was, we had to find the way down and I kept having to wait for my brother, or have him wait for me when I wasn't sure of the route. There was a lot of standing around waiting.

H: Again, how long did the second bivouac last?

M: Six hours. Through the hours of darkness.

H: You still had the space blankets?

M: No, not anymore. The bivy site was in the middle of a steep wall. The terrain had gotten much steeper. We just crouched on a ledge.

H: Recovery effect?

M: Zero. A night on the mountain in the freezing cold really drains you. And we were getting more and more dehydrated. But there was nothing else to do other than sit it out. Either you sit there and try to survive the night or you might as well jump off.

H: As soon as the moon was up, you continued down?

M: As soon as I thought it was possible. Up there it's easier to keep moving than it is to sit and do nothing, both psychologically and because of the cold. We waited until the impenetrable darkness had passed, then we kept climbing down. We no longer had any sense of time; hours, days, they all blended into one. Our only thought was that it would be light soon and we could get going again.

We kept climbing down, first in the moonlight, then in shadow. It was slow going. The terrain was difficult: icy rock, gullies, relatively steep. But we were no longer 7000 meters up; we were at 6000, then 5500. That makes a big difference. I also had the strange feeling I'd climbed the route before.

H: How would you describe your grasp on reality at that point?

M: At the bivouac, we both kept hearing running water. I even went for a look at one point, as I thought there must be a spring or something. But there's no spring up there. It was just a hallucination, wishful thinking caused by thirst. That delirium-induced state of altered perception became more and more frequent.

H: You were aware that you were hallucinating, though?

M: No. There were moments when I told myself, "That can't be right; there must be something wrong with me," but that kind of altered perception had now become part of my reality. There were a few sections on the descent that seemed familiar. I had the feeling I'd been there before, that I knew the route, and that the rest of the descent wasn't a problem. I told Günther, "We've been here before. I know the way."

H: How did he react?
M: Irritably at first. Then he said that I must be mistaken. But basically, he accepted my delusions.

H: Why were you so far ahead?
M: Because the routefinding required it. It wouldn't have been possible to get my brother down safely otherwise.

H: But there were times when you couldn't see him at all.
M: A mountain face is not a ski slope. The lower section of the descent route was like one big hanging glacier: unstable seracs, crevasses, humps and hollows, huge frozen waves of ice.

H: Why did you have to go so far ahead to find the way down?
M: Because the terrain was so confusing when viewed from above. The idea was to save Günther from having to climb back up. I often found myself in a dead end and had to retrace my steps, climb back up, and look for a better way down. Günther wouldn't have managed another step uphill.

H: What did the terrain look like at that point?
M: Concave, with a series of seracs dissecting the face above. Thick layers of ice everywhere, and where the face was steeper, there was a kind of incision where the ice broke off in huge slices, as big as skyscrapers. Sometimes that would happen every few minutes, then it would get quiet again—but for how long? It was imperative to get us out of the fall zone of those seracs as fast as possible. There was a constant fear of dying, a fear as palpable as the danger itself. Instinct took over, and I ran down to get out of the avalanche zone.

Meanwhile, Günther must have reached the dip at the foot of the steep section, as I couldn't see him. When he didn't arrive, I wasn't too worried at first. I consoled myself with the thought that he would be down soon, that it was no big deal; he was probably just having a rest. Why should anything have happened to him now?

H: You thought you were out of the danger zone?

M: He was still in the avalanche zone, but he only had a few hundred meters to go. Where I was, the avalanches would probably not have gotten me. I was totally exhausted. The relief hit me, and I just collapsed. But I was calm. I thought we were safe.

H: He didn't arrive. Why didn't you go back up?

M: Because I was dead tired. And because I told myself he couldn't have disappeared, he couldn't be dead.

H: How long did you wait for him?

M: I didn't wait. There was a second possible descent route from the lower, concave part of the face. To get a good view of that route, I had to get onto the lateral moraine over to the right. That was obviously where Günther would come down, I thought, since he hadn't come down the same way as me.

The main reason I came down the way I did was because it was hard snow, avalanche snow. The alternative descent was over scree and boulders, not dangerous terrain but awkward and strenuous, although there weren't any crevasses, which was an advantage. The question was, which was the easiest? I opted for the hard, compact snow. I would say I probably took the longer, more strenuous route, but it was easier terrain.

I persuaded myself that Günther would be coming soon perhaps because I didn't think I'd be able to make it back up again. After what we'd been through, I wasn't able to simply wander back up the way I'd come. I kept telling myself that my brother would be down soon. It was only after a while, when he still hadn't appeared, that I started to get anxious. From where I was, it looked like there was a way out of the valley down below. But first I needed to drink. I found a spring and just kept drinking and drinking.

H: When did you realize that something had gone wrong?

M: When I was back on the glacier, standing in the full glare of the sun and I heard the avalanches. I was safe from them now, but I was getting more and more worried about my brother.

H: What time was it when you realized he was no longer coming?

M: It was sometime in the morning.

H: The snow was soft?

M: Yes, I was sinking in when I walked back up. It was torture. With the exhaustion, the soft snow, and the burning sun, it took me several times longer than coming down had done just a short while earlier. I could only manage it one step at a time. One step, rest, then another step.

H: What was your mental state?

M: Total desperation.

H: Panic?

M: Not yet. I was still hoping I'd find my brother at the foot of the face and be able to help him. Of course, I did wonder if he might have fallen into a crevasse or if he'd been caught in an avalanche. There were so many "ifs."

H: Where did you find the strength to go back up again?

M: I had to find my brother.

H: How long had you been without food by then?

M: I'd had no real food for nearly four days.

H: Did you have any energy left at all?

M: No, I had no energy left. Everything went so slowly, every step was hard work, all that existed was pain.

Something like that is unimaginable for the layman. All he wants to know is how I could have left my brother and gone down ahead. Well, anyone would have done the same in my position, with those seracs ready to avalanche at any moment.

And now it was all about going back up there to find my brother. There was no alternative. It was the natural thing to do. I was following an inner compulsion.

H: How long did you spend back up on the face?
M: I went up and spent the whole afternoon searching for him, then I spent the night on the glacier.

H: You bivouacked for a third time?
M: I didn't bivouac; I just wandered around up there. I was going crazy, half-awake, half-asleep. I knew by then that my brother was dead, but I could still feel his presence.

H: How did you know that?
M: The fact that he must be dead was an obvious conclusion to draw. But my emotions told me otherwise. I had the constant feeling my brother was somewhere close by.

H: Could you hear him talking?
M: I heard him talking, and I heard him walking. When I walked, he followed along behind me. I also heard him shouting.

H: What did he say?
M: I went to where the shouts were coming from, but there was no one there. But Günther was there. All I had to do was turn around and I would see him, or so I thought. Reason told me that my brother was dead; there could be no other answer. But emotionally I was aware of his presence.

It went on like that the whole afternoon, the whole evening, the whole night long. I searched until morning, and then I went down to the spring. When the sun hit the face, I knew the next avalanches were coming.

H: Were you still able to mobilize any last reserves of energy?
M: Going downhill was all right. Going downhill is always all right. It's always much easier than going up, especially at altitude. The higher you go, the easier the descent is compared to the ascent. On the summit of Mount Everest, it's about 1:10, ten times less strenuous than climbing up.

H: You said you suddenly found yourself in a green meadow?
M: I'd been having hallucinations the previous day—a green field and people. And these images were now becoming more and more vivid, to the extent

that I could even see specific people. At first, I was sure they were our friends from base camp, who had come to help us. I knew the camp was on the other side of the mountain, but I had no idea how far away it was. I was sure the others would turn up there, though. They weren't obliged to, but it was the natural thing to do.

But in my deluded state it was completely different people that turned up—former climbing partners, famous mountaineers, my mother, even complete strangers. I saw horses, a horse with a rider, shepherds, a herd of cattle. The longer I spent on my own, the more of these apparitions there were. And I'd just spent another night up there on my own. I only headed down to the valley when I realized no one was coming. The valley was empty.

H: Were you absolutely certain by then that your brother was dead?
M: On the one hand, yes, when reason prevailed, but I still had the feeling he was there somewhere. I kept trying to ascertain where he was, then I'd realize again that he was dead. I looked everywhere for him, and with the search came the realization, the certainty, that my brother was dead. There was no other explanation. I'd have found him otherwise.

Theoretically, I could have found him dead, in which case it would have been easier to head back down on my own. The grief would have been the same, of course.

H: After the torture you have described, you yourself were more dead than alive. How did you cope with the thought that the end was near? What did you expect to find in that unknown valley?
M: I told myself I was going to die soon, that if I didn't find any locals I would starve. It was no big deal; it was just the way things were. I wrapped my spare clothing and the other stuff I had with me in my jacket, pushed my ice ax through it, slung the whole bundle over my shoulder, and headed down into the Diamir Valley.

After a few steps I went back and put one of my gaiters (blue and red, I think it was) on top of the big boulder I'd bivvied under, then placed a few stones on top of it. I was still hoping my companions would come, still hoping a search party would arrive in the valley, by helicopter or on foot. That's why I marked the bivy site. I didn't have paper or a pencil, so I couldn't write a note to explain what had happened. Then I set off down to the valley, staggering, crawling, lying down for a rest, then dragging myself to my feet again and taking another few steps. I had no idea how far it was. It was a miracle that I found the way down at all. There was no path, and I made the spontaneous decision to cross a glacier even though that was illogical.

I have often asked myself why I did that, why it was that I managed to find the only way down into the valley. But I could just make out some vague yak trails and saw some dung, and I ended up following the trail the herdsmen use in the summer to get to the grazing in the upper part of the valley. That's how I got down to the tree line. And that's where the people were.

H: The people who found you and then helped you down, who carried you on a wooden stretcher lashed together with yak-hair rope on the final stretch down to the road, and safety.

Baur, the expedition cameraman, now maintains that you had planned to make a complete traverse of Nanga Parbat all along. He says you wanted to establish a world record and you sacrificed your brother in the pursuit of this obsession.

M: That's just laughable. If that were the case, why didn't Baur radio down to base camp on June 28 and tell them to send someone over to the Diamir Valley right away, as that was the way the Messner brothers were coming down and that if they actually made it down alive, they'd be dead on their feet and need immediate assistance?

It was a sorry state of affairs. Of course the others would have come around to the Diamir Valley if they had known that we were going to turn up there, but nobody could have expected us to do that. All that talk about traversing the mountain was just wide-eyed imagining about the future; I never considered it as a real possibility.

H: Just idle talk to pass the time in the tent then?

M: Why on earth shouldn't you think about such things? The idea of a traverse was a topic for discussion. But a world record and fame? No, there was never a concrete plan to do anything like that. And anyway, the evening before I set off from the top camp, I told Baur to wait there for me until I got back. "Wait here until I get back," I said. It's in his diary and even in the expedition report, so it isn't just how I remember it. I didn't want them to go down; I wanted them there, for psychological support. I also asked them to take the 200-meter rope that was lying outside the tent and fix a section of the couloir, so the descent would be easier for me.

If I'd been planning to traverse the mountain, I'd have been clever enough to take some things with me: extra gear, a bit of money, my passport and ticket. And some food. I'd also have taken my bivy bag, something to drink, and maybe a stove as well, although not necessarily a tent, as it would have been too heavy. But I'd have taken the minimum amount of gear needed to survive; I wouldn't have wanted to risk dying up there.

The fact was, I only intended to go to the summit and come straight back down again. The red rocket meant bad weather was forecast, so what I was looking at was a quick dash to the top and back. That's why I asked Baur and my brother to wait for me. Otherwise, I'd have let Günther in on the secret and told him, "I'm going over the mountain and down the other side. If I'm not back in three hours, it means I'm not coming back this way and you two can head back down if you want."

H: Another expedition member, Max von Kienlin, says that when you finally got down to Gilgit you were totally exhausted and had no idea where your brother was. He says there was no talk of an avalanche and that you and he concocted this version of events.

M: Why should I have made anything up? Max von Kienlin, who was an expedition guest with no mountaineering experience whatsoever, merely advised me to stand by my "dead in an avalanche" statement. My brother was lost without a trace. I told Herrligkoffer about the tragedy first, then von Kienlin—exactly as it had been. Herrligkoffer knew all about it before von Kienlin even knew I was still alive. There was nothing to concoct.

H: The expedition leader, Herrligkoffer, later said that Günther died after the first bivouac; in other words, he never came down the Diamir Face.

M: Yes, that's what he said. Herrligkoffer was never there, of course, but he needed an excuse for his behavior—the red rocket, the failure to send a search party to the Diamir Valley, setting off on the journey home without us—so he spent a lifetime inventing justifications for his actions and allocating blame, in order to divert attention away from his own responsibilities. He kept coming up with new excuses.

H: He later presented you with a statement of events that he asked you to sign?
M: Yes. I was supposed to sign his made-up version of the events surrounding my brother's death—to lie, in other words.

H: What did the statement say?
M: The statement said that Günther died at the first bivouac and that I then descended the Diamir Face alone. After I was presented with that outrageous document, I told myself I wanted nothing more to do with the man. Ever. That's when all the disputes started, in 1970.

H: Why were you asked to sign it at all?
M: I don't know. Probably so I could stand convicted of planning the traverse of the mountain all along. How else could Herrligkoffer assuage his guilty conscience? Maybe he thought that there was no need to go looking for the Messner brother who hadn't stuck to the plan—and for his brother, who had acted out of personal ambition and had dreamed up such extravagances for his own ends, even less so.

H: Why didn't Herrligkoffer's team come looking for you in the Diamir Valley?
M: I can understand that. If Herrligkoffer hadn't accused me of sacrificing my brother to my own ambition, I would never have leveled my own accusations against him. Herrligkoffer, like the majority of the team, was pretty sure that we had both been killed. Understandably enough. They were all sad, and some of them really distressed, when we didn't turn up. Back at base camp, the team was naturally dejected; everyone was unhappy. (Kuen had a problem, for sure, when I turned up—after all, it meant he was no longer the sole all-conquering hero of the Rupal Face of Nanga Parbat.) But because they all thought that the Messner brothers were dead, they set off on their journey home. There was no chance of finding two dead bodies somewhere high up on the mountain, so sending out a search party would have made no sense.

H: Do you blame yourself now for Günther's death?
M: There is no one else to take the blame, only me. I bear full responsibility. That's why I don't understand why a few of the expedition members still keep trying to twist the story.

I also keep asking myself why it was that I survived and Günther didn't. When I set off for the summit, I was responsible only for myself; it was only my life that I was risking. That needs to be acknowledged. That solo summit bid was a bold move. When my brother followed, the scenario suddenly became totally different. I naturally felt responsible for him, and I knew that if we kept going we might find ourselves in a pretty tight spot. We kept going anyway.

Whether or not we made the right decisions is no longer relevant. If I'd been as experienced then as I am now, I'd have abandoned the attempt right after he joined me and headed back down to Camp 5. We could have taken a different route down the Diamir Face, too, but I didn't know that at the time, because it was all unplanned. After the summit we were forced to make a series of crucial decisions. We could either stay put and die, or risk the impossible. The lower down the mountain we got, the more risks we had to take. It was only our survival instinct that stopped us from giving up.

H: Did you have to go so far ahead on the descent?
M: Nobody would have done it any differently. If the two of you are moving at the same speed, you descend the mountain together, parallel to each other. But not if one of you is struggling and can't cope with any detours or having to go back uphill to find the best way down. In that case, it's up to the other guy to do the routefinding.

So I had to keep going on ahead, as far as was necessary to be certain we could get down the next bit safely. The situation was critical; it was a matter of life and death. There were only the two of us up there, and we had to make some big decisions. After the first bivouac between the Rupal Face and the Diamir Face, when we realized the others couldn't get to us, and beneath the seracs of the Diamir Face, we had to decide what to do.

H: You had seven toes amputated in the hospital in Innsbruck because of frost-bite. How did you cope with that?
M: I just had to resign myself to the situation. I soon got used to it. I had to learn to live with the phantom pain and other problems. Autogenic training helped. I still didn't give up climbing, though.

H: But at first you must surely have thought you'd never climb again?

M: I never actually thought that myself, but the first time I saw Herrligkoffer he took a look at my toes and told me, "You'll never be able to climb again." A normal person would never say anything like that in that situation, but for Herrligkoffer I guess it was important to tell me. He was a doctor; maybe he thought I'd be impressed by what he said.

H: You mean it was wishful thinking on his part?

M: In 1953, when Hermann Buhl came down from Nanga Parbat, Herrligkoffer behaved in a similar way. Buhl was the savior of that expedition, and after Buhl's success, Herrligkoffer, the expedition leader, decided to leave the remaining equipment from the Nanga Parbat climb in Pakistan so it could be used on Broad Peak the following year. Herrligkoffer told Buhl, "We're going to Broad Peak next year, my next eight-thousander. But you can't come; your feet are wrecked."

You have to understand Herrligkoffer's thought process here. The man had identified so closely with Buhl's summit success that it was as if he, Herrligkoffer, had gotten to the top himself. But it was Buhl who came back down from the summit, and that was when Herrligkoffer's inner conflict became obvious, even to Herrligkoffer. Buhl was Buhl—he really existed— so what was Herrligkoffer supposed to do with his own summit feelings? He had to marginalize Buhl, to act like he no longer existed.

H: How did you explain the tragedy to your parents?

M: I just told them everything.

H: And how did they react?

M: Differently.

H: What did your father say?

M: When we first met, Father asked me why I hadn't brought Günther with me, and I told him. Mother simply accepted it. She was the one who suffered most, for sure, but she also understood my situation.

H: Did your father blame you?

M: Of course. The tragedy was every bit as difficult for my parents to cope with as it was for my sister and my brothers; it was the first time anyone from our family had died. It was easier for me. I'd experienced the tragedy firsthand. I knew what had happened. For the others, it was all so very far away. Pakistan, Nanga Parbat, 8000 meters—it was hard for them to comprehend, hard to cope with.

H: You must have suffered terribly knowing that you had survived and your brother hadn't.

M: That feeling of guilt has remained with me to this day. It's part of the tragedy for every survivor. You end up thinking, "Why did I survive and not him? Why me?" The ideal scenario would have been if both of us had survived, but that's not what happened. It seemed so unfair; it still does.

H: Don't you find it odd that the tragedy on Nanga Parbat was not the end of your Himalayan adventures but just the beginning? Surely it would have been easier to say, "I'm giving up now. I never want to experience anything like that again."

M: This is the key point, the contradiction that nobody wants to understand when it comes to my mountaineering exploits. No climber ever wants to go through anything like that. But I am trapped between the desire never to experience that kind of thing again and the desire to have similarly intense experiences over and over again.

It is through resisting death that we humans experience what it is to be human. And it is in this seeming paradox that the most fundamental reasons for climbing mountains or seeking out extreme situations are to be found, whether it's the South Pole, the North Pole, the Gobi, K2, or Chang Tang. The secret lies in the fact that I can only have the most intense experiences when I push myself to the limits of what is possible. Obviously, when I'm doing that, I hope that I won't die and that nothing bad will happen to my partner, that everything will go well. I also know that if I just go for a bit of a hike, the experience will not be so intense.

There's a very good story about this, written by Eugen Guido Lammer. It takes place about a hundred years ago. He falls into a crevasse, and it's three days before he gets out again. During that time he experiences death, and hell. On arriving home, he experiences heaven. A feeling of having been born again. That is the answer.

H: So what is the message of the story?
M: It's all about not dying in potentially fatal situations. Gottfried Benn put it well when he said that mountaineering is about challenging death and then resisting it. Death has to be a possibility. The art of mountaineering lies in resisting it, in surviving.

I never want to have to endure an experience like Nanga Parbat again. But by the same token, I can't live without the experience of pushing things to the limit. The symptom of my disorder is defined by a lust for life that comes from putting my life at risk.

H: Was Nanga Parbat the most intense experience you have ever had in that respect?
M: Yes, for sure.

H: Because you were balancing for so long on the knife-edge between life and death?
M: Because the balancing act went on for so long, and because death was so close that I ended up taking it for granted. As I was going down the Diamir Valley and when I met up with the local people, I felt a bit better.

I eventually got to the stage where I couldn't walk any longer but I couldn't cope with being carried either. I had dysentery, so it was pretty desperate. So I scrounged some paper and a pencil and scribbled a note for one of the locals to take down into the valley, in the hope that he might be able to find a police station and organize a helicopter. My English wasn't so good, but between English, Italian, and Latin I managed to cobble together a message of sorts. The locals couldn't read, but they didn't have to; they knew what I meant. I finished the note by saying that I only had a day or two to live. Simple as that.

In other words, I was quite matter-of-fact about it; I took my imminent demise for granted. I had just a couple of days left to live, and then I'd be dead. It was the end, but not the end of the experiences I had.

H: What do you mean by that?
M: That knowing I was going to die wasn't a problem; it had become something that I just took for granted. At the same time I still wanted to reach

safety. The will to survive was still there, as was the desire to do everything I could to get down to the valley.

H: But surely you had already reached safety?
M: No. If I'd ended up stuck in Diamiroi, the dysentery would have killed me. I should never have drunk the dirty water there, even though the locals who gave it to me were able to stomach it.

The note I gave them never made it down to the valley. One of them took it and disappeared, but the helicopter never came. The messenger probably thought that the white man was going to die anyway, so he didn't need help anymore.

H: You describe it almost cheerfully. A positive experience?
M: It was a long time ago. And coming to terms with death is a pleasant, peaceful state of mind. Death is a fact. It's a part of life. It wasn't like I wanted to die, but I realized that death really wasn't much more than a last exhalation of breath. I perceived this fact with something like relief.

H: So death is something we should happily accept?
M: As long as the hope of staying alive is still there, the thought of dying is frightening. When all hope vanishes—as my experience on Nanga Parbat taught me—something redemptive comes over us, an acquiescence, an understanding of death. Then, finally, a slow sinking into death itself. No, dying is not so bad.

It is this experience—not dying itself—that you want to have again. I keep hoping to experience it again but without anything bad happening. It's got nothing to do with having a death wish; it's all about hungering after the experience. I'm not caught between the fear of death and a death wish but between the feeling of terror that I might die when I'm up there and the joy of having survived.

H: When did you embark on the first expedition to look for your brother?
M: As soon as I could walk properly again, in the autumn of 1971.

When I got back to the place where I'd lost my brother, it reopened all the old wounds. In the depression at the foot of the face, where all the avalanches funnel together, I searched for his body for days.

On the first day, right after arriving at base camp, I went up to the glacial basin at the foot of the face, hoping to find Günther's body. Just like that. Somewhere in the ice. It was late autumn, the snow had melted, and even the avalanche cones had shrunk. But I could find no trace of my brother. Eventually, I started searching the whole area. A glacier is constantly in motion, so his body might have been moved.

H: How big an area did you search, roughly?
M: Four to five square kilometers, I'd guess. To no avail.

H: Your partner at the time, Uschi Demeter, says you cried every night in the tent.
M: It all kept welling up inside me. The loss of my brother was like a phantom pain. And by then, of course, Herrligkoffer had also invented the story about our plan to traverse the mountain, which meant that searching for us had been unnecessary.

H: A few years ago, your former expedition colleagues Max von Kienlin and Hans Saler suggested in their books that you sacrificed your brother to your own ambition on Nanga Parbat and that you either left him up on the mountain or sent him back down the Rupal Face. In an interview with *Stern* in 2002 you said, "Sooner or later the remains [of Günther] will turn up and the matter will be resolved."
M: I knew that was the only proof that the people who doubted me would accept.

H: In a press conference Max von Kienlin maintained that if your brother really was ever found at the foot of the Diamir Face, your critics would be made to look stupid. Long before that, on July 20, 2000, to be precise, on an expedition organized by you, the mountain guide Hanspeter Eisendle found a human bone at 4400 meters at the foot of the Diamir Face.
M: It was a shinbone. My brother Hubert, a doctor, was on the expedition, and when he saw the bone, he said it couldn't have been Günther's because it was too big. I took the bone back with me anyway and kept it in my library in Juval.

H: Why did you wait three years before you sent it off for DNA analysis?
M: The body of the Pakistani climber I thought the bone had come from

was found elsewhere, which meant that the probability that I had Günther's bone had significantly increased.

H: After months of tests, the Innsbruck Medical University professor Dr. Richard Scheithauer came to the following conclusion: "The collective results of the classic forensic tests, and in particular the DNA analysis, show beyond a reasonable doubt that the bone that was examined originates from a brother of Reinhold and Hubert Messner." After the professor's findings, did any of your former expedition companions, who had accused you of sacrificing your brother to your own personal ambition, apologize to you?
M: No.

H: On Manaslu, your next eight-thousander, it all went wrong again. Why?
M: It really did all go wrong again. After that mountain I told myself that from then on, I'd do it all on my own. Every time I went away with other people, something went wrong.

H: Why did your companion Franz Jäger decide to go back to the camp on his own?
M: I don't know. Because he was tired and it was a long way to the summit maybe. Those 8000-meter peaks are hard work. The young lads probably think climbing at 8000 meters is a bit more demanding than in the Alps, but they have no idea just how tough it really was slogging up the mountain at that altitude with the equipment we had back then. Like Günther and me on Nanga Parbat—we had to find that out for ourselves, too.

On Manaslu the situation was exactly the opposite of how it had been on Nanga Parbat two years previously. Two of us set off together; one of us wanted to go back down to the camp. The only question is: was it justifiable for him to turn back and me to keep going up? Based on my experience, that was totally okay. It felt like it was okay at the time as well.

H: How far away was the camp?
M: The descent to the camp was certainly doable in one or two hours. Easy walking terrain. But I can only guess at what actually happened. Franz

Jäger must have reached the last camp long before the storm came in. Then he left the tent to look for me and got lost.

How is that possible? Well, in a whiteout you lose your bearings in an instant. And that was the worst snowstorm I have ever experienced.

H: You summited alone. What did it look like up there?
M: The summit is a little spike of rock, plastered in ice. After the first ascent in 1956, the Japanese hammered two rock pegs into the summit tooth and tied a flag to them. Big, long pegs. I took one with me.

H: What time were you on the summit?
M: Late afternoon. Too late, really. Not as late as on Nanga Parbat, but not before noon, which is what you should aim for, as a rule.

H: And how long did you stay up there?
M: Not long. I only had an hour or two to get back to the camp.

H: What made it the worst snowstorm you have ever experienced?
M: It snowed heavily, a meter of snow, and the wind was blowing spindrift everywhere. Then it got dark, pitch black.

H: You deliberately ripped your windproof suit to stop yourself from sliding off if you fell?
M: That's right. But the suit only offered flimsy protection anyway. The big problem was that I'd strayed off course. I'd lost my bearings.

H: You tore your beard out because you couldn't get any air?
M: It was all iced up. It was dark, and I couldn't see through my goggles anymore.

H: Was this another instance of pushing things to the limit, the experience that you so desired yet also feared?
M: No, in that situation I wasn't scared of dying, not at first anyway. I was just hoping to find the tent as soon as I could. I searched the big plateau, just kept going. Sometimes I was on ice, and sometimes I sank up to my chest in the snow. Spindrift and holes everywhere. And it was dark.

H: Visibility?

M: Practically zero. Eventually, I realized I was going around in circles. It was clear to me that I would have no chance of surviving if I simply stayed where I was. I would just freeze or die of thirst, I thought.

H: Was that when you heard the voice of your companion, Franz Jäger?

M: Yes. I thought he was in the tent, shouting my name. So I shouted his name, but then his voice kept coming from somewhere else. He kept shouting to me, and I kept shouting back. Maybe he'd left the tent.

The only thing that finally helped me find my way was my knowledge of the terrain and the weather. I knew that I was on a big snow plateau and that this was to the north of the ridge where our tent was pitched. The storm was coming from the south; I noticed that when I was on the summit. So I told myself I had to walk into the wind, which meant I was heading in a southerly direction. I got to the ridge, the snow ridge where the tent had to be, and I knew that I would find the tent somewhere along that ridge. And that's what I managed to do.

But Franz wasn't there. Presumably, he had gotten back to the tent much earlier, made himself some tea or whatever, and then left the tent to look for me. I got there and saw the tent and shouted, "I'm here!" And someone came out, and I thought it was Franz Jäger, that it couldn't be anyone else. But Andi Schlick and Horst Fankhauser were there. Andi said to me, "You're totally hypothermic; you stay in the tent, and Horst and I will go and look for Franz. We'll have him back here soon." So I waited in the tent. Later, when they still hadn't returned, I kept shouting, but I didn't leave the tent.

H: They didn't come back?

M: No. I couldn't understand it. But it turned out they'd lost their sense of direction within fifteen minutes and couldn't find their way back to the tent.

H: Where did they end up?

M: In a snow hole under a kind of snowdrift. They dug themselves a hole and let themselves get snowed in.

H: How long did they sit it out?

M: Until morning. Andi Schlick kept insisting that they should try to find the tent again, so they left the snow hole but they couldn't find the tent and had to dig another snow hole. At some stage Andi Schlick left the second snow hole and never came back. He just disappeared. Fankhauser couldn't stop him. Schlick was probably hallucinating and thought the tent was close by.

H: Was there an argument?
M: No. Horst said, "Just stay here; it's no use. We have to wait for daylight, then we'll find the tent." But Schlick went anyway.

H: When did you get the feeling that neither Jäger nor Schlick were coming back?
M: When Fankhauser came back in the morning, it was already light. The storm was over. We talked, cooked, and went out to see where the others were. But there was no sign of them, nothing. A person, even a dead body, would show up as a dark speck against the white background, but there was nothing, nothing at all—just snow, snow, and more snow. We spent a few hours searching for them on the plateau. Then, about midday, we started down. There was no other choice. The weather forecast was not good.

H: Were the bodies ever recovered?
M: There was a report much later that two bodies had been found on the plateau, but they couldn't be identified.

H: What was the decisive factor in your decision to go down?
M: The deterioration in the weather. Stormy weather and snow on the way. The avalanche danger between Camp 4, where we were, and the camp below was already enormously high, and it was due to increase. The climbers below us, especially my friends Bulle Oelz and Wolfi Nairz, the expedition leader, were urging us to get down to where it was safe as quickly as we could.

Also, it would have been impossible for the two of us to search the huge summit plateau with avalanche probes on our own. We were hoping to see a black speck or something sticking up out of the snow, but unfortunately, there was nothing.

H: How far was it to the lower camp?
M: It was a good 800 meters lower. It was evening by the time we got there.

H: What was the mood like when you arrived?

M: Bad, of course. Two very good friends had just disappeared.

H: Were there any recriminations?

M: This time the recriminations only came from outsiders, from Herrligkoffer and from Kuen again. The Herrligkoffer brigade was on Everest that same year. They didn't get very far, and once again there was trouble on the trip. Of course, the know-it-alls again said that what we did was irresponsible.

H: What did Herrligkoffer say?

M: "You never leave anyone behind on the mountain." He had plenty to say. However, what he forgot to tell people was the fact that on the 1953 Nanga Parbat expedition he ordered all the camps below the last camp to be cleared when Buhl set off on his summit attempt.

H: Why did he do that?

M: Because he had forbidden Buhl and the rest of them from going for the summit. Imagine that. They clear all the camps, so the lead climbers are stuck up there on their own. If anything had happened to Buhl, no one would have been able to help him.

H: Isn't there a rule that you should never let anyone go back down on their own?

M: What do you mean by a rule? Who makes the rules? Looking at it another way, you could say that all of the great successes have mostly been achieved against the rules. There is also a "rule" that says you shouldn't solo routes. So what? If you've got a climbing partner and a rope and can belay each other, then great. But I can still do things differently if I want to.

Climbing mountains is not a sport, it's not a game, and it's definitely not a religion. I go into a dangerous area, a life-threatening world, and I take responsibility for myself and my partners. And they do the same. We all try our best to survive. But it's dangerous, and there are some situations when it's not possible to survive. If you don't understand this, or you don't accept it, you shouldn't climb at the limit. No one is forcing you to, after all. You should do it of your own free will—accepting responsibility for yourself in the sure knowledge that you are exposing yourself to risk—or not at all.

H: But surely the greatest individual mistake that you can be accused of making is letting one of your climbing partners go back down alone?
M: The mistake, if there was one, was climbing Manaslu in the first place. There are no rules. If circumstances permit, you can let someone go back down on their own. If that wasn't the case, I would never set off with a partner.

H: What did Kuen accuse you of?
M: Nothing concrete. Those guys kept making their moral accusations, but they were always pretty generous where their own behavior was concerned. Two years later Kuen killed himself. I never commented on that, and it's not my business to comment on it now. I never said suicide is immoral, and thus unacceptable. It was his decision and his alone.

H: By 1972 you had climbed two big mountains in the Himalayas. The result: three dead.
M: That was hard for me to deal with. The other thing was, if I'd been on my own it wouldn't have happened. If I'd summited Nanga Parbat alone, I would have gotten back down to the last camp, and the whole thing would have turned out differently. I would probably have made it up and back down again the same day in spite of everything. On Manaslu, if I'd set off on a solo summit attempt, I wouldn't have had anyone else to worry about. But Jäger wanted to come with me at first, and I couldn't just exclude him. Nor did he have any misgivings about going back on his own.

It was these experiences that strengthened my resolve to try things on my own in future. When I'm soloing, I am responsible only for myself.

H: Did these tragedies prey on your mind at all?
M: The feeling of guilt, having survived when others died, naturally preys on your mind. I still ask myself what the others would be doing now. Günther, Andi, Franz—who would they be now?

H: Did you suffer from depression?

M: No, I didn't suffer from depression as such. Depression is not the right term for it. It's more a feeling of despair and grief. Later on, there's a kind of survival guilt; you feel guilty that you are still alive.

H: How long does the grief and despair last?

M: For years. Having said that, it was less intense after Manaslu than after Nanga Parbat. That's probably because I was much closer to my brother than to my climbing partners on Manaslu, who I'd only really gotten to know on the mountain.

I didn't know Schlick and Jäger very well before the trip; all I knew was that they were good climbers. On Manaslu we were a team, we became friends. Nairz was a very companionable expedition leader. We were together on the mountain, but the close friendship only developed later. We were kindred spirits, and after the trip we all grew closer. We all had feelings of grief. Horst Fankhauser and I had had a pretty traumatic experience, and the others reacted with sympathy rather than recriminations. After all, they were climbers, too. Unlike Herrligkoffer.

H: And it was then that you decided to go it alone?

M: Yes. I decided to do an eight-thousander on my own, to attempt it at least. In 1973 I tried to solo Nanga Parbat, but I failed relatively low down because I couldn't come to terms with the dangers, the fear, and the loneliness. I felt so lost and lonely that I turned back. I wasn't able to cope with that degree of exposure on my own. I could no longer think clearly. I felt like I was going to pieces.

H: It sounds like lunacy. Previously, you had only ever been to the Himalayas on huge expeditions, and suddenly you decide to climb Nanga Parbat alone, with only a rucksack?

M: Yes, and right from the bottom. Technically it wouldn't have been a problem, I could have coped with the objective difficulties on my own. The problem was emotional. Because I had no one with me to talk things through with. A problem shared is a problem halved—"This is hard; what do you think, should we go back?" I was lost, at the mercy of my own loneliness. I didn't have the mental strength to keep climbing on my own.

H: How would you describe that inner turmoil? Were you having panic attacks?
M: It wasn't the fear of falling or the fear of getting stuck higher up the route, just the fear of being alone on that big mountain. I was gripped by a feeling of absolute doom. When I looked up and saw the size of it all, the endless expanse, I didn't dare take another step. I was scared—scared of losing my mind. There was a feeling of infinite loneliness, like a black hole, as deep as Nanga Parbat was high.

H: What qualities do you look for in a climbing partner on these big mountains?
M: He or she must be reliable, enthusiastic, and strong. You are going into the wilderness together, after all. If one of you decides to turn back, that's not a problem. It isn't the summit that's important; it's the shared experience. Often it's only a few hours from the last camp to the summit, plus a few hours for the descent. You can manage that on your own. But setting off alone from the bottom of the mountain is a totally different thing. You miss having someone else with you, someone you can rely on, someone to share the fear and the joy with, someone who needs you as much as you need them.

When you are on your own for days on end, the climbing gets progressively harder to do because the loneliness and the fear increase. The difficulty and effort are compounded by being on your own. Being out of your social comfort zone is one thing; losing all human contact is another thing entirely. It wears you down. I was mentally unable to cope with that level of self-reliance, even though I was physically very fit.

H: How far did you get?
M: I got to about 6000 meters. It was my first attempt at soloing an 8000-meter peak. In 1974 I tried another eight-thousander, and again I failed. By then, I had also started climbing new routes on other big mountains in South America and looking for extremely difficult routes on mountains all over the world.

The next big step in high-altitude mountaineering came in 1975 on Gasherbrum I [Hidden Peak], when I was able to do what I'd initially wanted to try solo—but with a partner, Peter Habeler. It was a revolutionary step, doing hard routes on the highest mountains in the world with a minimum of equipment: acclimatizing on other mountains first and then climbing the route and descending. Up to that point, all the ascents of 8000-meter peaks had been made using high camps and fixed ropes, and a minimum of two tons of expedition equipment. We only had 200 kilos in total.

After our success on Gasherbrum I, I developed a plan to attempt Everest in the same style. It was clear to me that this would only be possible if I could climb the mountain without oxygen. If I needed supplementary oxygen, an alpine-style ascent would be out of the question. In the 1970s the oxygen equipment for Everest weighed 50 kilos on its own—a handicap of seven bottles per man.

After climbing Mount Everest without a mask, it was clear that all the mountains in the world could be climbed in the style I had adopted, with minimal equipment and limited means but with total commitment and a higher risk factor. As long as you had the experience and skill and were fast.

H: What did you have in your rucksack on the Everest climb?
M: Everything you need to survive for ten days. It weighed 15 to 20 kilos. Tent, mat, sleeping bag, food, fuel, spare clothing, of course—three spare pairs of socks, for example, to reduce the risk of frostbite. Everything else I wore—boots and several layers of clothing. Plus several pairs of gloves and a stove.

H: You each carried a stove?
M: No, we only took one stove with us. On longer trips, like in Antarctica, you take two stoves in case one breaks. We also packed cooking equipment, and gas cartridges, of course. I generally count on about half a kilo of gas and about 600 to 800 grams of food per day for two people—packet soups, bacon or salami, but not too much fat. It's harder to digest food at high altitudes. The stomach doesn't get enough oxygen either.

H: Was there not a chorus of outrage at your idea of setting off without bottled oxygen?
M: The main criticism came from scientists, who thought my objective was not achievable. The physiologists and doctors were clearly of the opinion that my experiment would not work.

H: You then took a plane ride over Everest without using oxygen. How did that go?
M: In 1977, a year before the attempt on Everest, I flew over the mountain

in a Pilatus Porter piloted by Emil Wick. It was very cold. I was well accli-matized, however, as I'd just got back from Dhaulagiri. I got in the plane in Kathmandu, and within two hours we were circling over the summit of Everest, me without an oxygen mask. I'd been told I would lose conscious-ness, but that didn't happen. I never put the oxygen mask on at all. I took some photographs, got some really nice shots of Mount Everest. At about 7500 meters I had a bit of a crisis, felt a bit clumsy, but higher up I was fine again. I knew then that it was on; the only question was whether I'd be able to summon the concentration needed to climb at that height.

In 1978 Peter [Habeler] and I decided to give it a shot, but we went with a big group rather than as a two-man team. We all helped with the organi-zation of the expedition, and my friend Nairz was the expedition leader. It was a cumbersome expedition and it was relatively expensive, too—they all were in those days. Peter and I paid our share of the costs and were given the opportunity to do what we wanted on the mountain, to operate independently of the others.

H: Why was it that Peter Habeler then decided he didn't want to attempt the summit without oxygen?
M: That's what he wanted to do. It was understandable, really, since without it the chances were not that good. But the others in the group had already paired off and the order they would climb in had been decided, so Peter's proposal was vetoed.

H: Why were the others angry?
M: They weren't angry; they just weren't flexible enough. There were two different projects in the group. Peter and I were the team climbing without bottled oxygen. Then all of a sudden Peter wanted to join the others, which meant he needed oxygen equipment after all. But whose bottles was he supposed to use, and who was he going to climb with?

Then there was the logistical problem of fitting in another ascent with oxygen. It was the others, not me, who took exception to the fact that he enjoyed certain privileges from the start and then suddenly he wanted to attempt the summit in the easiest way possible, using the others' bottles.

H: What kind of privileges?

M: The others had to follow a strict time plan and climb in a predetermined order. We had first try, and then the others would have their chance. If they didn't manage it, they had to go to the back of the line. Cutting in line didn't go over too well.

H: You said that compared to Nanga Parbat, Everest is an easy mountain. Why is that?

M: Logistically, climbing Everest isn't easy. It's a very long way. And there are also some technically demanding sections, right at the start, dangerous sections. But like I said, at the start. If these difficulties came at the end, climbing Everest would be hellish. On the Rupal Face of Nanga Parbat, the major difficulties are right at the top.

H: The major difficulties on Everest are in the Khumbu Icefall, aren't they?

M: Yes, we had to install ladders there for the Sherpas. Nowadays there is a special team of Sherpas whose job it is to pre-equip the icefall with fixed ropes and ladders. The climbing teams then follow, and they have to pay to use them.

H: On your first summit attempt, you were caught in a horrific storm.

M: Yes, there were three of us. Peter had gone back down, so it was just me and two Sherpas up there. We had two big tents and a tiny one as a reserve, a kind of mini-tent for extreme emergencies.

A bad storm came in. But it wasn't the snowstorm we'd feared, just wind—forty-eight hours of nonstop gales. We didn't get meters of spindrift, just wind, wind, and more wind. The tiny tent held out in the end.

H: You made a solemn promise that you would forget about Everest if you survived that storm. Why didn't you keep it?

M: We couldn't use the stove, so we couldn't eat or drink. The Sherpas were just as worried as I was. One of them was an uncommonly helpful man, and very determined; the other wasn't frightened exactly, just distraught. He curled up in his sleeping bag in a corner of the tent and lay there as if he were dead. From time to time he even acted as if he were dying, moaning and blaming me for everything.

The wind was so strong that we couldn't even go outside to pee, so we had to pee in our water bottles. Afterward, that one Sherpa stated publicly that I'd given him urine to drink. Frozen urine? I don't rule out the possibility that there was some lying around, but it would have been frozen solid. We had nothing to drink the whole time; everything was frozen, and we couldn't use the stove.

H: What was wrong with the stove?
M: The wind came through the tent seams and blew it out. It was a desperate situation. It was cold, and we thought we might die up there. The others couldn't get up to us, and if the storm had lasted three more days we would have been dead. In situations like that, you often tell yourself you'll never do it again if you can just get away with it that one last time.

H: I read that to save weight, Peter Habeler cut the matches in half.
M: We took the bare minimum of gear with us. We even cut the long straps off our rucksacks to lighten the load. In the end, with various little tricks we maybe saved half a kilo in total. We didn't cut the matches in half, though—of course not. That's not important anyway.

The sad thing about the whole affair is the Habeler book you are referring to, which wasn't written by Peter but by a Munich journalist who wasn't there. He also wrote that Peter and I had agreed that if one of us wasn't able to keep going, he'd be left behind. And he came out with all the old clichés that we climbers find so annoying, because they are simply not true. The idea that either of us would simply keep going to the summit even if the other guy was dying is ridiculous. There was never any such agreement. There never is. We didn't even speak about it.

If something unforeseen happens, you naturally go back down together if at all possible. Each of us would instinctively try to rescue the other. That is something you don't even have to discuss with your climbing partner. If your partner is dead or can't be rescued, then you try to get yourself off the mountain safely.

H: After the summit you became snow-blind on the descent. Was that due to an error of judgment?
M: An English film team had given us two cameras, one for Peter and one

for me. At the last moment, Peter refused to take his camera with him. It was too much extra weight, he said. I'd promised to film the last bit to the summit, however, so I'd packed mine. I had an obligation, and an ambition, to document the summit bid.

H: What did the camera have to do with you becoming snow-blind?
M: Plenty. I kept taking my goggles off to film and take photographs. There was a strong wind blowing, and they kept icing up. I wouldn't have been able to see properly otherwise. Finally, I took them off and kept them off—a stupid mistake. The blood supply to the brain is not as good at altitude, so you forget things. I often used to forget to put my goggles on.

The sun was really intense above 8000 meters, and that, plus the strong UV light and the ice crystals whipping through the air, damaged my retinas. My eyes were burning, but there were no other symptoms of snow blindness at first. That only started when I was back at the camp, luckily; otherwise, I'd never have gotten down. Peter had already set off.

H: What is it like to be snow-blind?
M: You can't see a thing, nothing at all. Your eyes are painful and watery. When it really started, my eyes were watering constantly and the pain got worse and worse. It was like I'd had sand rubbed in my eyes. The next morning I could see well enough again to make out shapes.

H: How did you manage to get down the mountain like that?
M: It went all right. I just stumbled down after Peter. Naturally, he kept having to wait for me, and he told me which way I needed to go. By the time we got down to the Western Cwm, I could see okay again, and the next day, on the descent to base camp, the snow blindness was completely gone.

H: Habeler writes in his book that you lay in the tent crying and that you said, "Don't leave me here alone, Peter!"
M: That's a bit of an exaggeration. I actually said it would be better for us to make the descent together. Peter was my partner.

H: It sounds like you were scared he might leave you on your own, like there really was some kind of ominous pact that you'd be left behind if you couldn't make it down.

M: That's the perfidious thing about this story. The journalist I mentioned before obviously thought up a plot that sounded good to him. But firstly, there was no need to ask Peter to help me, as he would obviously have done that anyway. Secondly, if he had left me, it wouldn't have shown him in the best light, would it?

There were three other guys at the last camp, too, but I wanted to descend with Peter Habeler rather than with two Sherpas and the British cameraman, Eric Jones. If Peter had said, "I'm in a hurry and I need to go down now, at my own pace. I can see fine and you can't," the only thing I could have said was, "Okay, I'll follow you down later with the others." It was just taken for granted that Peter would go down with me. I repeat: it was taken for granted. If I'm up there with my partner and he's snow-blind, I slow down and wait for him. I still don't understand why Peter didn't have that passage taken out of the book like he promised he would.

H: You had a falling out over it, didn't you?

M: Yes. Later that year, 1978, I was on the way back from my solo ascent of Nanga Parbat, on a flight from Frankfurt to Munich. I was flipping through some magazines when a German journalist and climber named Hermann Magerer gave me a copy of the Habeler book. I started reading it and stopped at the bit about leaving your climbing partner behind. Instead of going home, I drove straight to the Zillertal to see Peter. I told him, "That passage in the book is wrong. It's stupid and it could lead to misunderstandings. I'd always thought our partnership was different than that. What you wrote was out of line." He defended himself by saying that he hadn't written the book himself, so I said, "Yes, but you must have read it." He said he'd get it corrected in the next edition, but when the time came the publisher prevented that from happening. Business was more important.

For me, that was the end of our climbing partnership. Unfortunately, it also meant I'd lost a friend. But I didn't want to climb mountains with a partner I couldn't trust or, even worse, one who would leave me to die and go for the summit on his own. How could you misrepresent one of the core values of mountaineering in that way? Or tolerate a statement like that just because it would sell well? When business becomes more important than the facts, that's where I draw the line. We had a real conversation about it four years later, but we never climbed together again. The trust had gone.

It was a shame, and it was disappointing, because for ten years we had defined fast-and-light mountaineering. We did the north face of the Eiger in ten hours. The previous fastest time was eighteen hours. We were the first team to climb the eight-thousander Gasherbrum I without high camps, ropes, or high-altitude porters—and without supplementary oxygen, of course—and we did it via a difficult new route. It wasn't doing Everest without a mask that drove a wedge between us but a stupid comment made by Peter, or by his ghostwriter. The end of our partnership was not marked by a groundbreaking climb but by a stupid comment made by a nonclimber.

H: Would you say that Everest without oxygen was your greatest success at the time and the climb that most captured the public's imagination?
M: Yes, the Everest ascent was the first climb I did that had worldwide resonance. They were even talking about it in the American magazines. Why did it cause such a sensation? Only because in the run-up to the climb, everyone had doubted it was possible. The book about the expedition became my most successful book, even though it was the worst one I've ever done.

H: Why was that?
M: I edited it together from tape-recorded diary entries in two weeks. Eighteen or twenty cassettes I'd recorded during the expedition were typed up. Half the stuff was unusable; the other half I cobbled together in a few nights. In spite of this, it became a best seller. It was translated into several languages and sold half a million copies in Germany alone. It meant I was able to finance further expeditions.

H: You had achieved a new level of fame, so to speak?
M: Yes, but it wasn't thanks to a new style of climbing. Everest without oxygen was less revolutionary in climbing terms than the Gasherbrum climb. But now I had the possibility of earning more money. I was getting more inquiries—so many that I could never have satisfied all of them—and was able to increase my fees as a result. My profession was now starting

to pay off financially, and I landed a few advertising contracts. From that moment on, I no longer had any problems financing my projects, and that was the prerequisite for my actions.

H: Were the advertising contracts very lucrative then?
M: A five-year contract for boots or rucksacks brought in around 50,000 deutsche marks [roughly US$25,000]; not much, but enough to finance two or three expeditions a year. Climbers don't earn much compared to other athletes—around a thousandth of the amount a Formula 1 driver gets, for example. Climbing trips cost a lot of money, however, and a top climber will probably want to do at least three big trips or expeditions a year. Giving lectures and writing books is actually counterproductive, as it means you lose your fitness.

Nowadays it is even harder to make a living, as the best time to be a mountaineer, financially speaking, is over. There is more competition, so the pie, which is small anyway, has to be divided up between more people. Also, because the top climbers don't seem to understand how to support each other, they are always trying to talk down each other's achievements. And as mountaineering on prepared trails has become more fashionable, classic alpinism has lost out.

H: You failed on your first solo attempt on Nanga Parbat because you couldn't cope with the loneliness. What gave you the confidence to go back and try the mountain again?
M: I had split up with my wife, so privately I was in a totally new situation. Secondly, after my experience on Everest, I knew for sure that I was well able to cope with being at 8850 meters. Nanga Parbat is 8100 meters, so the problem was comparatively smaller, so to speak. I figured on climbing it in three or four days maximum and one day for the descent. I hadn't planned on getting snowed in.

H: Do you climb better when you have just split up with your wife?
M: We'd been together for a long time; we were a strong couple. We'd done pretty much everything together. Uschi Demeter had also been my secretary. I learned a lot from her. She had more experience of the world than me, and

a different approach to it. I came from the narrow confines of a South Tyrol hill-farming community; she was from the sophisticated city of Munich. For me, she epitomized the big, wide world, and she gave me access to it.

H: And why did she split up with you?
M. There were several reasons, for sure. Eventually, my no-compromise attitude to climbing and the fact that the only thing I ever thought about was my next expedition became too much for her. She increasingly felt more like a hanger-on than my partner and found that she no longer enjoyed coming with me on expeditions. It's not nice sitting around at base camp knowing that your husband is at 8000 meters and might not be coming back.

She wanted to travel to other places, too, to visit exhibitions and go to the theater. Instead of bringing us closer together, my successes—Gasherbrum I, Everest without oxygen—created a distance between us. I had become a public figure. Finally, we separated. I was too focused on my own objectives, working on new boot designs, for example, that would perform better at high altitude than the conventional leather boots. Now I was a single man again, a free agent.

H: Boots instead of marriage. Was it worth it? What did the new boots look like?
M: They were made of plastic and closed-cell foam rather than leather, loden wool, and fabric. My new boots stayed dry. When a leather boot gets wet, it can freeze into a solid, heavy lump of ice. This causes frostbite. My invention was warm and light, and it became the boot of choice for high-altitude climbing for the next twenty years.

Nowadays there are even lighter boots made from Kevlar, Gore-Tex, and carbon fiber, but it's still all about materials that don't absorb water. Since this type of boot became available, there have been significantly fewer instances of frostbite. Previously, every other climber on expeditions to 8000-meter peaks lost their toes.

H: How long did it take you to solo Nanga Parbat?
M: Three days up, after acclimatizing on a neighboring peak. My last bivouac was at 7600 meters. I'd only planned on climbing during the day, as I needed to be able to see properly on the really difficult sections, particularly

on the first part of the route, where I might have to dodge avalanches or spindrift coming down from above. Climbing the Diamir Face of Nanga Parbat at night would be lethal.

H: What emergency supplies did you take with you?
M: There's no room for emergencies on a solo ascent like that. There's nobody to call in an emergency. If I had injured myself badly, there would have been nobody to help, because nobody would have known I was in difficulty. A broken leg would have been fatal.

H: If there were two of you, surely you could splint the leg?
M: Yes, if there were two of you. Depending on the terrain, you could maybe get the injured person off, even lower him down on a rope. But a rescue is only possible in very rare cases, and self-rescue is generally out of the question. With fixed ropes and twenty people on the mountain to help, you could even recover a half-dead climber, but it's infinitely more difficult to do if there are only one or two of you; in fact, it's more or less impossible.

H: What time did you reach the summit? Last time you didn't get there until five o'clock in the afternoon.
M: This time I had no problem with the timing. There was a lot of snow on the last section leading to the summit. Not good snow either, but I knew that getting back down was only a matter of a few hours. I got back to the tent at 7600 meters without any major problems. I hadn't overexerted myself, keeping to the rock wherever possible instead of wading through deep snow. I climbed as fast as I could comfortably manage, without pushing myself to the limit.

I spent quite a long time on the summit, looking around and taking photos. I knew that no one would believe I'd soloed it if there was no photographic evidence. It was the first solo ascent of an eight-thousander. Then I went down to the tent.

H: Did you leave anything up at the summit?
M: A little capsule with a note inside.

H: You must have derived particular satisfaction from having climbed the mountain alone.

M: The real satisfaction came from having overcome my fears, my doubts, and my weaknesses. I hadn't had to make compromises or argue with anyone. There was no logistical plan of ascent, I just kept climbing, making my plan hour by hour. It was the perfect trip.

I was autonomous, responsible only for myself. I planned and executed it all on my own and succeeded against all expectations. In mountaineering circles, that climb achieved the greatest recognition of any climb I have done. Two new routes on Nanga Parbat—solo! Then there was the solo ascent of Everest, which I envisaged as a kind of finale for my high-altitude mountaineering.

H: And a liberation from it?

M: Yes, but it was more than that. I had always seen Everest without oxygen and Nanga Parbat solo as a way of gaining experience for a solo attempt on Everest. Soloing Everest was going to be the icing on the cake in my career as a high-altitude mountaineer. However, because I couldn't get a permit for a solo ascent of Everest at first, I tried K2 in 1979. K2 was thought to be the most difficult eight-thousander.

H: Before that, however, you had another narrow escape to contend with, on the descent from Nanga Parbat. You were snowed in at 7600 meters?

M: For forty-eight hours the weather was so bad that I couldn't see a thing. There was no chance of finding the way down. If you can't see, you can't get your bearings. Even though I knew the mountain well, I just had to sit it out. I also knew that the bad weather can last a long time, up to two weeks sometimes.

H: And how long could you have lasted up there with so little oxygen?

M: As long as I could drink, three weeks or so. Without gas, one week maximum. I had enough gas to last me for a week. I kept looking out of the tent, lying there half-asleep, then staring out into the mist and snow. Take a look outside, nothing, then back to sleep.

H: Were you scared this time?
M: No, I tried to push all those feelings away.

H: What did you think about?
M: Nowadays I'd probably think about my museum, how to set it all up or something. I came up with the idea for my mini-museum in Sulden in a tent, during a storm in Greenland.

H: You imagined yourself in another world?
M: Yes, that's the trick I've always used. It's interesting, like viewing your life as an outside observer. You have to learn to think about something completely different, something challenging, to stop yourself from going crazy in critical situations like that. If all you think about is how long it will be before you get rescued—ten minutes, ten hours, or ten days—you start to panic.

Everything is easier to cope with in tiny doses, even happiness. I think about all kinds of things. I let my imagination roam free. That solves a lot of problems.

H: Nevertheless, the waiting must be agony.
M: Time passes very slowly up there. A day lasts a week. At base camp I've always got something to read, I can do some writing, listen to music, or chat with the others. We used to spend the whole day playing cards to pass the time. None of that is possible in the high camps. How are you going to carry a book up there, in addition to all the food and gear? A book weighs about the same as one day's rations, so that would mean one day less to play with, a narrower survival margin.

H: So what did you think about when you were snowed in up there?
M: I don't remember now. In Antarctica I thought about Juval. I rebuilt my castle and turned it into a museum. In my head. By the time I got home, I had a plan all worked out—down to the last millimeter. In Greenland I came up with the concept for the little museum of curiosities in Sulden am Ortler.

H: In other words, to enable you to cope with it all, during the expedition you think about the next expedition or the next project?

M: More than just the next project. Ideally, it should be something that has nothing to do with the mountains. You're lying in your tent, and you know that your tent is the only safe place to be; you won't be swept away and you won't freeze to death. But sooner or later, the waiting has to end; otherwise, you'll die of thirst.

On Nanga Parbat I had only one chance of getting out of that situation. I needed the clouds to clear enough to allow me to see right down into the valley so I could get my bearings, with hopefully a few hours of daylight left to get down in. I tried to keep myself busy. I didn't try to convince myself that I wanted to stay alive. I just let my mind wander and thought up new projects so I had a reason to stay alive.

H: And then you climbed all the way down the Diamir Face, a total height loss of 3000 meters?

M: Yes, early in the morning on the fifth day. After two nights in the tent, there was a short break in the clouds and I could see which direction to go and memorize the way down. I decided to take a line straight down a gully, as I knew I couldn't miss the route even in the mist. I climbed down on rock and ice—hard, steep ice. This was the only way down, and it was dangerous, so I had to move fast.

H: Where did the danger lie?

M: Chunks of ice could break off the big seracs at any moment. The lower I got, the greater the risk of avalanches and falling ice. I was lucky. In just six hours I got down to a height of 5000 meters. Once I was finally out of the danger zone, I was able to take my time. The weather remained bad for weeks. You don't always get a gap in the clouds like that, so I was doubly fortunate.

H: Your next trip was to K2, a mountain that many regard as the most difficult eight-thousander, a killer mountain where the weather is always bad.

M: K2 gets better weather than Nanga Parbat. The topography of K2 is different, too, so there are fewer avalanches, although everywhere is very exposed to the wind. K2 is steeper than Mount Everest and saves its most treacherous sections for the upper part of the climb. Many climbers have

High-altitude mountaineering: on the north face of Chamlang. In the background is Makalu, the fifth-highest mountain in the world (1981).

perished up there—frozen, fallen to their deaths, or died of dehydration, pulmonary edema, or apathy.

H: How did you approach K2?

M: We arrived in the country at a time of political instability, unfortunately. It was 1979, and ex-president [Zulfikar] Bhutto was due to be hanged. For weeks everything in the country was at a standstill. We were stuck in Islamabad. We lost a lot of time, which meant our possibilities were limited. We didn't have enough time to do the route we'd planned. The south face looked too dangerous, so we switched to the Abruzzi Spur. Climbing this as a small team with only three camps was a really bold thing to do.

H: What does the summit look like?

M: Completely white, a snow peak, a beautiful summit. It was late by the time we topped out, six o' clock in the evening. We were crawling on all fours to get through the snow on the uppermost part. The snow was so deep that we had to posthole a trench to get up there. It was very time-consuming.

H: How did your companion, Michl Dacher, do?

M: Michl climbed like a machine. We took turns breaking trail. One of us would do a bit, then the other would take over. In spite of the avalanche danger, we got up there safely and back down to the last camp, where we used the last of the light to anchor the tent. By then, the first stars were coming out. The weather stayed fine during the night—it was just very cold—but the next morning we were sitting in clouds, impenetrable clouds.

H: You described your solo ascent of Everest as the icing on the cake of your mountaineering career. Why then did you go on to climb the rest of the eight-thousanders?

M: When I visited Tibet for the first time, on the 1980 Everest expedition, it was like another world—the Tibetan culture, the wide expanse of the high plateau. I had found a totally new challenge. I thought it would be interesting to walk across the Tibetan Plateau, but I could only go back to Tibet with a climbing permit, so in 1981 I went to Shishapangma. I had a few other ideas up my sleeve as well: the "triple crown" of eight-thousanders, traverses of 8000-meter peaks, that kind of thing!

H: When did you complete all fourteen eight-thousanders?
M: In 1986, with Makalu and Lhotse, which I climbed as part of a larger expedition. I had achieved the last of my self-defined objectives on the 8000-meter peaks. A new chapter of my life could now begin.

My sponsors had other ideas, however. They wanted me to stick with being "a mountaineer." I wanted to go to Antarctica, and that's what I did, in spite of the opposition from my sponsors and the climbing scene in general. My sponsors didn't want an explorer, they wanted an active mountaineer to advertise their products. So I had to reorganize the fund-raising.

H: Where did the south face of Lhotse fit in?
M: That came after completing the eight-thousanders, as a kind of epilogue. I'd already failed on it once, back in 1975. In 1989 I tried it with an international team and we failed again, like everyone else before us.

H: What makes it harder than the Rupal Face on Nanga Parbat?
M: It's steeper and more technically difficult, but not as dangerous. The Soviets climbed it in 1990. It was a huge effort, with thirty people, several kilometers of fixed ropes—and oxygen, naturally. I tried it in 1989 with four good climbers and two helpers.

This time, I was only the expedition leader, responsible for the organization and logistics. That was the first time in my life that I went on a trip with no ambitions of summiting. Right from the start, I said I wouldn't go to the summit. I was forty-five years old and probably no longer capable of climbing such a difficult face, so I helped the others as far as the last camp. The others, however—all excellent alpinists from South Tyrol, France, and Poland—failed to commit to the hardest section of the route. They all wanted to summit, but none of them wanted to do the crucial trail-breaking work. Because I was just responsible for the logistics, I told myself I should step back and leave it to the others. Obviously, they were all thinking the same thing!

It's often the case that the bravest climbers set off first from the last camp but don't make it to the summit because it's too far or the climbing is too difficult. Sometimes it's the next team that manages it, in the tracks of the guys who failed ahead of them. On the 1989 Lhotse south face trip, that's what they all wanted to do. They went up to the last camp in teams of two and just sat there waiting for someone to seize the initiative. I found the whole thing amusing and annoying at the same time.

I'd seen that kind of behavior several times since I started high-altitude mountaineering in 1970. Generally speaking, the ones who later feel they were outsmarted simply failed to take their chance when it was there. If you break trail to the summit, the risks are greater, but never in my life had I experienced the situation where climbers at the last camp argued about who should go first. The others were always pleased that someone was prepared to make the first summit push. On that 1989 trip, no one could summon the courage to go for it.

We failed on the south face of Lhotse, but in my mind's eye I was already in Antarctica. A new phase of my life had begun.

CHAPTER III
INTO THE ENDLESS
WILDERNESS

1986–2004

*Airlines have allowed at best for an
almost absurd reduction in travel time, but not
in distances. Those remain as immense as before. Let
us not forget: The line of flight is only a line, not a
road. From a physiognomic viewpoint,
we are pedestrians and runners.*

—Christoph Ransmayr

DESERTS AND LOVE

My wife and children have brought stability to my life. I have not become sedentary, however. No one had ever completed the traverse of the Antarctic continent on foot, no one had walked the length of Greenland from north to south, and no one had even considered trekking across the Gobi Desert, a distance of 2000 kilometers, in a single push. So I thought I'd try.

Again and again, I kept getting drawn to the outer limits of civilization, where once more I had to learn to see the invisible, to cope with the remoteness and the exposure, to live with extremes of cold and heat, in the same way that I had learned as a child in the Dolomites to instinctively sense loose rock. And once more I was guided by the indigenous people—Eskimos in Greenland and the Tibetans in the Chang Tang Plateau, the Uighur at the edge of the Takla Makan Desert, and the last of the original inhabitants of Patagonia. These guardians of our planet became my mentors. So, too, did the Arctic night, the fog, and the open water of the Arctic Ocean. I had given up and turned back there once before. Another failure.

But I wanted to try again. We quickly agreed that the summer months were a better time for a trans-Arctic expedition than March and April. What we were unsure of was how we were going to get onto the centuries-old central ice mass and how and where we were going to make landfall in Canada.

Victor Serov, who had spent many summer months in research stations on the drift ice, confirmed that in July and August the terrain consisted of ice and freshwater lakes and that the temperatures were pleasant. The problem was that the open water channels don't freeze over after May, and it snows a lot. It's often quite foggy in the summer, too.

Victor Boyarsky explained the problems they had on the 1995 trans-Arctic dogsled expedition and told us about boats that could be pulled over the ice like sleds. We agreed on the following strategy: start at the end of April, beginning of May from Siberia, with two boats loaded with provisions and fuel for a hundred days. The plan was to wait at Cape Arkticheskiy—for several weeks if necessary—for a favorable wind and thick ice floes, and to drift with the wind northward. Provisions could be bought locally. In July and August we wanted to be in the middle of the Arctic.

At that time of year, the ice barriers have melted and it is possible to sled sail. We knew that the last hundred kilometers would be hard going. But we figured it must be possible to head westward in a big arc and reach Ward Hunt Island, even if it meant paddling. The tactics were: two boats with 150 kilos each—we'd hunt seals if necessary— two men pulling the sleds in turn, with a rest day every third day; and a third man to go out in front on his rest day to find the best route. Team members: Victor Boyarsky, Hubert, and Reinhold Messner. Logistics and organization: Victor Serov.

The Arctic Ocean crossing never happened, unfortunately, and later that year I fell off the wall of Juval Castle and broke my foot, which ruled out any further attempts for a while. It was just another failure I had to deal with. Failure itself is not important. It's what happens immediately after that counts—the inner feelings, the turmoil and self-doubt—and how you deal with it. It can mean a new start, an opportunity to experience your limitations and to grow as a result. My mind-set and my attitudes have changed over the years, and this is largely due to my frequent failures. I might not have mellowed much, but I have certainly become more resilient.

It is through failure that we experience our limitations. And it is for this reason that failure is a more powerful experience than success. When you get to the summit, all that means is that you've climbed the mountain, nothing more. By achieving your objective, the objective ceases to exist. When you fail, the objective remains. The despair may follow, the wish to understand the failure, the attempt to comprehend your own limits. I've certainly failed often enough. Even as a little boy. But those who never experienced failure in their younger years may later fail to understand the feeling of despair as a message, a realization of their own limitations, and instead perceive their failure as hopelessness.

Yes, to fail well requires practice, and you need to take little steps at first. The man who fails for the first time and allows it to destroy him is like the climber in the crevasse who does nothing to help himself, who does not use his aggression, his courage, and his strength to counter the despair. I did not keep taking on new challenges because I am more ambitious than other people; maybe it was because in failure I saw sufficient reason to risk another attempt.

H: You were running out of high mountains to climb. You had completed all the challenges you set for yourself. Did you have the feeling that there was nothing major left to do?

M: That's right! I had to invent a new challenge for myself.

H: You were getting older, too, and you realized that people like [Hans] Kammerlander, for example . . .

M: Were more skillful. Yes.

H: He was more skillful, and he was faster, stronger, and fitter. A reason for you to bow out?

M: Subconsciously, that played a part, for sure, but what was much more important for me was to find something I could get enthusiastic about, something I could identify 100 percent with. After eighteen ascents of 8000-meter peaks and thirty expeditions to the eight-thousanders, I couldn't really get that excited about those high peaks anymore. The worry, the fear, and the excitement I felt at the prospect of stepping into a totally alien world, like the first time on Nanga Parbat, was no longer there.

H: A romantic concept.

M: In 1986, when I trekked through eastern Tibet, I realized that kind of trip was also hard work; it was tough going, and not as easily manageable as the eight-thousanders logistically. On each of my eight-thousander expeditions, I had a huge pile of photos and information about the route to work with, so I knew how to go about it, but on a big walk like that, you can't predict and plan for every eventuality, let alone carry all the provisions and fuel you need for five months.

H: You are known as a perfectionist. What prompted you to expose yourself to that level of uncertainty?

M: I've always looked for problems that haven't been solved yet. Sometimes

I've invented projects myself, and sometimes I've taken them from other people, stole their ideas, if you like.

H: Isn't it much more boring walking across a plateau than climbing a big mountain?
M: Yes, that's what I thought at first.

H: So what is the appeal of a trek like that?
M: You don't have a well-stocked base camp you can return to, where you can recover or pick up some supplies. You're on the move all the time, reacting to events as they occur and adjusting your plans hour by hour, day by day. A desert trip is less plannable than climbing a mountain, and you are on your own for longer.

H: So a big walk has different qualities?
M: You perceive the world differently as a pedestrian than you do from an airplane or when you're climbing. You set off and you know that the next place you can buy food is 200 or 400 kilometers away. That means you have to eat what you can find locally. Even more importantly, you have to get along with the locals—like the nomads I came across in the wilderness of Tibet who threatened to set their dogs on me. That kind of thing can happen. It was all a lot more difficult than I'd thought it would be.

On the eastern Tibet trip, I had a heavy pack full of food, first-aid equipment, and clothing—everything I needed to survive, in other words. It was a relief when I met up with a yak caravan, even though communication was rudimentary—just hand gestures and a few words of Tibetan I'd picked up. The first thing they did was throw my rucksack on a yak. I stayed with them for four or five days, ate with them, and drank tea with them. I gave them a few little items as payment for the food and the help carrying my pack. But they were not part of my planning and logistics and I wasn't dependent on them. We finally went our separate ways, and I went on alone. I would never have gone home again otherwise.

H: What were you looking for on that trek?
M: I viewed it as a research trip. On the ancient trail of the Sherpas. These mountain people live in the Everest region now, but they originally came

from eastern Tibet. They migrated in the sixteenth century—twenty thousand people and as many yaks headed west in a mass migration—but no one really knows why. Eventually, they reached Nepal. This story was researched and reported by a German ethnologist in the 1960s. It was a mystery I found fascinating, and I wanted to trace the migration in my own way. I wanted to see if the story might be true. So I retraced the exact path, following the route the Sherpas took.

H: Did you have a map, or did you just go by the story you'd read?
M: I had maps, but not detailed maps.

H: What did the locals give you to eat?
M: *Sha kampo*, dried meat. And *zampa*, a kind of barley flour mixed with yak butter and tea. It's really tasty. I ate only what the locals ate.

H: Did you have any problems with sickness and diarrhea?
M: I've got a very robust stomach. I eat everything. And I can go hungry for three days or so as well, if need be.

H: How long did the trek across Tibet take you?
M: Months. It was an ill-starred trip as well. We didn't have a permit, so we had to keep a low profile at first. Then our Tibetan interpreter ran off. He was scared that the Chinese would catch him. He had fled to Nepal in 1959, and it was the first time he'd been back to his old homeland. Then my partner, Sabine, and I went in separate directions. She went east, and I headed west, following the old Sherpa trail over mountains and through valleys—no roads, no railways, no infrastructure at all, and often no real paths.

For part of the route, I was on paths that were only occasionally used by yak caravans. In some places there was snow on the ground and I couldn't see which way to go, as no one had been that way for weeks. Sometimes all I could see was a single set of footprints, which was pretty exciting, as you obviously wonder whose tracks they are and whether you can follow them. One time, after I'd been walking for days and had just crossed a pass, I came across a man who was prepared to carry my rucksack to the next pass in exchange for money. When we got there, he wanted to go back the way we'd come, so I had to carry it down the other side.

Then I pushed on alone, into the endless wilderness of the high plateau. A single set of tracks in the snow took me into the bottom of a broad green valley. It was late, and I was tired and hungry, and away in the distance I could see a nomad tent. A big black tent. I didn't hesitate; I just headed for the tent, stopped about 200 meters away from it, and gave a signal. But the people looked like they were getting ready to set their dogs on me—big black Tibetan mastiffs—so I backed off, keeping my eyes on the dogs the whole time, and went on my way. I didn't get anything to eat that day.

H: Do you take it personally when you get such a harsh rejection?
M: When I'm in foreign places like that, I have to accept that people might see me as an intruder, an enemy even. I have no right to make demands. No one is obliged to give me an enthusiastic welcome. It can happen, of course, and you can often enjoy great hospitality once you've been accepted and if you are prepared to be generous, but it should not be taken for granted.

H: Do money or gifts help?
M: Of course. It would be ideal if I could pay for every meal with a Swiss Army knife or a watch, but it would be impossible to carry that many of them around with me.

H: What kind of image of humanity do you get traveling around in places like that?
M: People help people as long as they feel safe, especially in the wilderness. Mistrust is not a weapon; it is part of our instinctive defense mechanism. And the locals naturally want to benefit, too. Experience has taught me never to expect assistance without giving something in return. That's the way it was when I came down from Nanga Parbat, battered and bruised, in 1970. The first people who found me were really kind, but they also had themselves to think of.

When I went back there in 2003, I met the man who'd been the first to spot me. He was seventy-three years old. In the end I asked him why they'd left me lying outside under a tree with just a thin blanket instead of taking me into one of their huts. He explained to me that they were sure I was going to die anyway, because I was so cold and exhausted. The next day I tried to get them to carry me, but they wouldn't. They walked along next to me. It was only in the afternoon when I couldn't go any farther that they agreed to carry me and they made it clear I should pay them for it, in kind.

H: How would you interpret their behavior now?
M: The fact that they were skeptical and thought there was no sense helping is understandable when you consider how they live. There's no doctor and no hospital anywhere near. If anyone gets seriously ill up there, they die. The people out there are not Good Samaritans by nature. They help as long as it makes sense to do so; they are objective observers. They will give you something to eat if you give them something in return.

If you turn up in eastern Tibet with no food and no money, it is entirely possible that people will say, "That's your problem." Anyone crazy enough to go to eastern Tibet with no food and no means of payment only has himself to blame for his situation.

H: Granting hospitality without immediate reward is regarded as a cultural asset. Were you really accorded such hospitality so rarely?
M: My situation as an adventurer is different than that of a guest. In the United States and in Europe, hospitality is a matter of course. It's like that in all wealthy societies, with a few exceptions. In the high country of New Guinea, however, things are very different.

In self-sufficient societies, for example, I can't expect to turn up and the people to say, "Great—nice to see you!" On the contrary, they will be thinking, "What's this foreigner doing walking through our place?" Adventure as an end in itself is a crazy notion for many people out there. The Tibetan nomads just try to survive out there with their yaks their whole lives long.

H: So you don't expect basic human kindness; you proceed on the assumption that people will scrutinize you first?
M: Why shouldn't they view me with skepticism? Like the farmers in Villnöss when I was a young lad. "What are those teachers' kids doing, off climbing all the time? Haven't they got anything better to do? If they want to get some exercise, they should be chopping wood or fetching hay down from the meadows. Climbing rocks! What's that all about?" That's what they used to say.

The South Tyrol farmers remained skeptical of what I did for a long time. "That Messner is a madman," they used to say. "He's really got a screw loose." They only acknowledged me once I achieved financial success. That's understandable. It's damned hard making ends meet as a hill farmer. They don't know where I get my money from, but it's a case of "he's got his own business," and they respect that.

In short: I've met plenty of skeptical people who don't necessarily respect the activities that others are up to but will help if trust is established and if there is a trade-off. That's human nature. Once they are certain that they're not going to be robbed, cheated, or lied to, they'll give you everything you need.

H: What did the landscape and culture of Tibet give to you?

M: For me, the landscape of Tibet is second to none. The air is clean and clear—it feels lighter—and you can see a long way. There are no garish colors, just pastel tones. For me, Tibet is the most beautiful country in the world, alongside South Tyrol, Patagonia, and Bhutan. I would really like to live there.

If it were an independent country, I would move there for a while and restore the derelict monastery of Shegar Chode. It's perched on a 400-meter-high almost-vertical cliff, about 5000 meters above sea level, with desertlike scenery all around.

H: Did you have a fundamental religious experience in Tibet, of a type that was unattainable for you in Europe?

M: No, but my attitude to life and my personal principles were reaffirmed there. I didn't become a Tibetan Buddhist; I remained a pantheist.

H: Did you encounter any behavior or attitudes that might serve as an example to us Europeans?

M: Compassion. More in the sense of empathy than sympathy. It's not a matter of Christian charity, either, which is something that is taken for granted in our cultural environment, even though it is often no more than lip service, the overblown emotionalism of moral values that have been learned by heart.

Compassion has little to do with morals or rules. The people of Tibet do not behave in an effusively cordial manner, and they are not intentionally altruistic. They are humane by nature.

H: Could you give a few examples?
M: Well, there was this one time in the Ténéré, but it could just as easily have happened in Tibet. I was there with my son in December 2003, and we had joined a caravan. One evening I cooked a spaghetti sauce, with my own mix of spices and lots of butter. I gave half of it to the men sitting around the campfire. They accepted the food in a matter-of-fact way that required no further comment.

There was no "Great, we'll give you something in return"; they simply tasted it to see if they liked it. They didn't waste any words thanking me; they just ate and said nothing. There was a common bond, a feeling of togetherness. There's nothing nicer than that.

H: It sounds like very bad manners!
M: They ate it all up, and they were happy with it.

H: You mean the fact that they ate it shows that they liked it, and that is preferable to someone standing up and saying that it tasted good?
M: Yes. Here in the West there's a lot of false emotion and playacting in social rituals; it can often feel very contrived. That's not the case with many of the people in Tibet. I prefer their direct approach—even when someone comes up to you and says, "You can't come through here or I'll hit you with this stick," he will doubtless have his reasons.

H: Can you think of any more examples?
M: Ang Dorje Sherpa on Kangchenjunga—and the matter-of-fact way he helped me down the mountain after the tents got destroyed. I was sick and had problems seeing. Ang Dorje understood the situation, even though his English was rudimentary, and he reacted to it.

H: You chose Antarctica for your next big trip, an unpopulated region with the ice and the wind your constant companions. Why was that?

M: I had the feeling that those wide-open spaces might provide the ultimate adventure. No one had ever crossed the Antarctic continent on foot before. It was a new challenge for me.

H: You once said that Everest was child's play compared to the Antarctic. A slight exaggeration perhaps?

M: Yes and no. In the Antarctic there's no danger of falling off, no avalanches, and enough oxygen. There are crevasses, but no big, steep faces.

As for the rest, the dimensions are inconceivable; everything is very much bigger. The storms really take their toll, and you can't just say, "Tomorrow I'm going to head back." Once you are in the middle of the expedition, you can't simply give up like you can on a mountain.

H: [Robert Falcon] Scott referred to the Antarctic as "this silent, windswept immensity."

M: A bland statement. When I was preparing for the expedition, I figured it would be extremely cold, and I tried to adjust the list of supplies accordingly. I thought about what we would eat—how many kilos per day—and how many kilos of fuel we'd need. I knew it was a long way—about as far as from Berlin to Moscow and back, maybe farther—and that it would be cold the whole time, thirty to forty degrees below zero, with frequent high winds and snow.

What I didn't know—and could never have imagined, fortunately—was just how strenuous it is pulling a sled like that in such cold temperatures. It doesn't slide. The snow is like sand. It's a real grind pulling a heavy sled across rough, windswept terrain.

H: Why is the snow so much worse?

M: Because it's so dry. The sled and the skis run slower when it's that cold, so making progress is harder work. But you are lost without a sled in the Antarctic. It's your home, your life-support system. Without it you'd be dead. So you just have to get on with it, even though it's a real burden. At a temperature of minus fifteen degrees, the sled runs quite well, and on ice it's even better, but every step is still hard work. In those conditions it's hard to find your rhythm.

The trick is to empty your mind and not to think, "Well, that's one more step done, just ten million to go," but it's not easy to do. My partner, Arved Fuchs, had that problem. After trekking to the North Pole a few months before, he found it hard to get back into the right mind-set again. Arved had reached the North Pole, albeit with a lot of air support, but now he had the hard work to deal with all over again. On the North Pole trip he'd had food drops almost every week. Maybe he thought the effort involved on the Antarctic trip would be similar. But when you're counting down the months right from the start, you're already doomed to failure.

On expeditions like that, I take it one day at a time; I think about that day's stage, then I drift off into fantasyland. I watch the light, which is often very beautiful. There are shades of grey, green, and yellow, depending on where the sun is and what the weather is like. When the weather turns bad, everything is grey. In a whiteout it's all dark, so dark that you can't wear goggles even if there's a storm blowing. But even then, you have to concentrate hard, shut out tomorrow and yesterday, and focus on the here and now. It's like being on another planet.

H: How did you cope with the cold?
M: The cold isn't a big problem. You're always plastered in ice with icicles on your beard, but you are always on the move. Or in your tent. You cope with the cold by walking or by lying in the tent in your sleeping bag. Walk, walk, walk, short rest so you don't get too cold, then off you go again.

H: How heavy was your sled?
M: When we set off, more than a hundred kilos. A lot more.

H: What happens when you get into a sastrugi field with a load like that?
M: Sometimes it just wasn't possible to pull the sled on your own and there was no chance of making any progress on skis, so we only managed 5 or 6 kilometers in a day.

Toward the end of their horror trip, Scott and his team only managed 2 kilometers a day. Two kilometers! It was fifty degrees below zero, and the next big food depot—One Ton Depot, with a ton of provisions and fuel—was 18 kilometers away. There were enough provisions there to last them the whole winter. If they'd managed to get there, they might well have survived—just dug in for the winter, built a big igloo, and spent the time sleeping, cooking, and keeping warm. But they were so tired and under-nourished that they could only manage 2 kilometers a day—and not even that much at the end. They knew that they didn't have enough food or strength left for those 18 kilometers. They died at their last camp.

In the final phase of our trek, we sometimes did 40 to 45 kilometers a day, but we were doing it in the "warm" season and we had reserves of fuel and food, whereas Scott was there in the winter. When you are so exhausted that you can't even make a mug of tea and the reserves are so low that you can't keep warm at night, you won't even manage 2 kilo-meters a day. The distance ahead of you seems endless, the problems too great to cope with.

If I'd known at the start of the Antarctica trip that we'd have to endure such hardship for ninety days, I'd have despaired. But after a week it got easier, and after six weeks we'd done a third of the distance. When we got to the Pole, I knew we'd make it.

H: Those heavy sleds must have felt like instruments of torture. Did the man-hauling leave you with cuts and bruises?
M: The problem is, pulling the sled puts a huge amount of pressure on the soles of your feet, so you can easily get blisters. I didn't, luckily.

H: But your partner suffered badly?
M: Arved Fuchs got terrible blisters. But he hardly ever complained; he gritted his teeth and carried on. You can't just stop where you are and call in sick, or the expedition is over. We had to keep going. The caravan must always keep moving; that's its raison d'être. There's nothing else you can do.

But the way Fuchs saw it, there was a tacit understanding that he would only go as far as the South Pole. That, plus the fact that he wanted to look after his feet, is why he only wanted to do short stages at first.

H: He didn't want to go any farther than the Pole? Did you discuss the matter with him?

M: Yes. We started talking about it after a couple of weeks. We were moving too slowly, you see.

H: What was his take on it?
M: He didn't think we had any chance of going the full distance. Besides being too slow, the snow was too bad. He was convinced we wouldn't get any farther than the Pole, which was why he wanted to make sure he made it that far at least. From there we could be flown out, he said. End of story. Just like on his North Pole trip, where he and the rest of the international team were also flown out, after six airdrops.

H: Could you relate to that?
M: Not really. Logistically, I'd planned the whole thing as a trans-Antarctica expedition. I'd raised the money, lots of money, and worked for months on the project. I was still of the opinion that we could do it. I wanted to stick to what we'd agreed: one day at a time, at least 30 kilometers per day.

H: How would you describe your relationship with Fuchs?
M: There was no tension between us on the trip. He wanted to go at his pace, and I wanted to go at mine, but there was only one solution really: we had to keep going.

H: You went at different speeds, and sometimes you were an hour ahead of him. Weren't you surprised at how patient he was?
M: Some people walk faster than others. That's just a fact—it's neither positive or negative. Everyone has their own ideal walking pace, the pace at which they get least tired. If I have to walk behind someone who is not going at my ideal pace, I get tired faster, much faster than I do if I can walk on at my own speed.

Fuchs and I had totally different walking speeds. I didn't intend to walk faster; I just went at my preferred pace. If I'd slowed down and walked behind Fuchs the whole time, I'd have tired more quickly and we wouldn't have gotten as far each day.

H: Still, it must surely be maddening when one guy is always out in front?
M: Yes, that's true as well.

H: It must have been really infuriating for him.

M: For sure, but what was I supposed to do? Give up with him at the Pole?

H: Did Fuchs show his anger?

M: No, not really.

H: Did you talk to each other?

M: We talked every day, whenever we were together—during rest breaks, in the tent. We only argued once.

H: When was that?

M: About 700 kilometers from the end, at the Gateway on the Beardmore Glacier. He wanted to pack it in there. He didn't want to move as fast as I thought necessary, right from the start. During the first phase, we weren't managing to maintain the speed we needed to travel to cross the Antarctic in the time we had available. So I protested. He said there was no point grinding it out, as we weren't going to make it anyway.

At first, I came up with the idea of increasing my pace as front man. We'd agreed to walk for six hours a day during that first phase. He didn't want to do any more than that. I didn't think that was enough, so I went ahead and did the six hours at my pace to cover as much distance as possible in the time. Fuchs needed seven and a half hours to my six, not counting breaks, and in the hour and a half I spent waiting for him, I got cold. That wasn't good. For either of us. And we only managed around 20 kilometers, give or take a few, in six hours anyway.

So I suggested reducing the weight of his sled and to keep reducing it until we were moving at the same speed. There was one condition: Arved had to stick to the seven and a half hours he needed to cover my daily distance, and I went ahead a bit faster than before. The first time we used this method we managed 30 kilometers in a day.

H: But Fuchs's feet must have been totally wrecked, really badly blistered, surely?

M: The fluid in the blisters leaks out after a while, and new skin forms underneath. Walking was painful for Arved, for sure. The fact that he was able to endure the pain is proof of his capacity for suffering. But we wanted to cross the Antarctic continent, not go for a stroll.

I admit that I probably behaved a bit cruelly at times, but what I will not put up with is the fact that after the trip he turned the whole thing around and acted like the big hero of the Antarctic. That annoyed me. I took the lead for the whole distance, apart from 5 kilometers or so, to ensure that the expedition ended successfully. Maybe Arved went ahead for 10 kilometers, but even that wasn't a hundredth of the distance, and it isn't really important anyway. Or it wouldn't have been, if Arved's manager hadn't played it up and issued those false statements after the event.

We set off in November, and I knew we had to be at the Pole by Christmas if we wanted to get to McMurdo by mid-February. If we'd only reached the Pole in January, we wouldn't have been able to continue, as we'd have been getting into winter. That's why I kept pushing Arved and telling him we had to get to the Pole by the end of the year. He said it wasn't possible, but we made it anyway. I admire him for that. And with those feet! If he'd packed it all in, I'd have had to give up as well. Arved could navigate, and he had more experience on the ice than me. It was all relatively amicable in the end.

I devised a little psychological ploy to urge him on to the Pole. I told him if we took fewer rest days and walked for eight or nine hours a day, we'd reach the Pole by the end of the year. And no one had ever managed to walk to both poles in a single year. I coaxed him along by telling him he'd be the first to achieve it; he'd been to the North Pole that May, and now he was battling on to the South Pole. Arved cursed a bit, called me an overachiever, then he canceled rest days and put in the extra hours. We reached the South Pole on New Year's Eve.

His achievement now stands in every record book—and every Fuchs book—the first man to reach both poles on foot within a single year. But I wasn't being ambitious on his behalf. I had an ulterior motive. I knew if we got to the South Pole by the end of the year, we could make it to the coast. And that's what we did. There was another crisis a couple of weeks before we reached the coast when we were getting worryingly slow again, but that was it really.

H: Some of Fuchs's people presented a different version of events, claiming that Fuchs took the lead and pulled you along.

M: Arved's manager thought he could put a spin on it, yes.

Fuchs did the navigating, and normally the navigator goes ahead. But Fuchs gave me the data each day and I went on ahead, using my compass, which I had around my neck, to check the course. Day in, day out. I didn't take the lead to show I was faster. I did it to save the expedition. I kept having to force the pace.

I would have preferred it if my partner had set the pace, if he'd said, "Right, today we are going to do eight to ten hours. We'll use the sails as long as the wind keeps blowing," or "We'll keep going all day and all night if we have to." But Arved wouldn't have been able to cope with that. That's why I never criticized him for it. And like I said, it's not important. If Holger Hansen hadn't come out with his completely opposite version of events, I'd have kept quiet about such minor matters. I just kept on pushing, coaxing, forcing the pace. We had to make the best of each day and see that we finished the job in spite of everything. And in the end we managed it.

It was only in the final phase that I got really impatient. After we reached the Pole, we started to get too slow again. Why? Because we weren't able to use the sails as much, because it got extremely cold, and because there were several glaciers to cross. Finally, at the Gateway, we had our only real argument. By then, we were just 700 kilometers away from the coast with flat ice and snow ahead of us. And we only had two weeks left. Two weeks for 700 kilometers. Fuchs said, "This is where the mainland ends. We've got the sea below us now." And with that, the Antarctic Transversale expedition was over for him.

But not for me. Fuchs said, "I'm staying here. I'm not going any farther." I said, "I'm carrying on to McMurdo. We can't stay here." We even ended up shouting at each other, and finally I said, "You stay here then. They'll come and get you. Eventually. I'm going to keep going on my own. I'll take my sled and half of the food and fuel. You can have the tent. I don't need it; I'll sleep in the sled. I'm going to the coast."

It was perfectly possible to get there by the middle of February, a realistic proposition. Why should he stay there? I'd expected too much of him, maybe. I'd argued against taking rests. I never slowed down; I just kept on forcing the pace.

H: Was that another of your psychological tricks?
M: No. I knew we could do it and that Arved was just being stubborn. That was understandable. And so we set off again.

After a day or two, the difficulties ended—no more glaciers, no crevasses, just storm-force winds from time to time. It was a long way. A very long way. But it wasn't so far that we had no hope of making it.

H: Were you finally able to use your sails?
M: No. The terrain was ideal, but the bad weather made using the sails impossible. We had to walk it—ten, sometimes twelve, hours a day. And we were hungry. The food was rationed. We lost a lot of weight and eventually suffered muscular atrophy.

Toward the end, we were setting off early and doing two or three hours straight, without a rest. I would let Arved get to within a hundred meters or so of where I was waiting for him, and then I'd keep going. I stayed a short way ahead the whole time. He complained a bit, but he stuck with it. I still find it remarkable that he managed to tough it out. A superb achievement!

H: Stony silence?
M: Yes, but what can you do. And in the end we made it. We arrived at the coast on February 13.

H: Nevertheless, it wasn't the start of a great friendship, was it?
M: No, but the partnership served its purpose, right to the end. It's important to note that at the time, I didn't know how to navigate. Initially, I had invited a couple of climbers to come on the trip. The finances were all in place. The idea was for the other guy to learn how to navigate and thus take some of the responsibility.

On big trips it's important to share the responsibility with your partner. He has to feel like he's irreplaceable, indispensable. But none of my climbing partners was prepared to go. It was too boring for some of them, too far for others. And hardly anyone had that much time.

The Fuchs/Messner team was the right one for the job. I still stand by that today. We were the ideal team in spite of our differences, or maybe because of them.

H: Because you rubbed each other up the wrong way?
M: Arved was slow and steady, and I maybe provided the necessary impetus. He was the brakeman and I was the musher, if you like. If we'd gone at his

pace, we'd never have made it, and if it had just been me forcing the pace unchecked all the time, we'd probably have worn ourselves out too quickly. It was the sum total of our characteristics and attitudes that made us successful. We actually complemented each other really well.

H: The ultimate common-purpose partnership.
M: It would have been easier if we'd been friends, particularly afterward. But we were a good team, with a common purpose. I'm glad I had that experience. I still like Arved Fuchs, without his manager in tow. As an expedition partner, he is considerate, calm, and tough. He knows his job. He didn't always walk fast enough, but I've never held that against him.

H: How long has it been since you last spoke to him?
M: Oh, years.

H: Which of your former partners are you still on speaking terms with?
M: I get together with Peter Habeler quite frequently. We're older now.

H: And Fuchs?
M: There were no problems until we got back home. We made a few appearances together in Europe. Then I read in the newspaper *Bild* that Messner had been dragged across Antarctica by Fuchs and that everything that had appeared in *Spiegel* magazine was lies and deception. Was I supposed to reply to that? I ended the contact there and then.

H: Which fundamental new experiences did you gain from the Antarctica trip?
M: Probably the most important thing was finding a new challenge. And the next new challenge was to be the North Pole—a similar expedition, but the difference was that I wanted to do it with no air support whatsoever. That meant I needed a new partner, a doctor preferably. You start getting a little fragile at fifty, not quite as steady on your feet.

My brother Hubert is a doctor, so I told him to ask around and see if he could find a colleague who had the time, the skills, and the endurance to do the trip with me. After just three weeks Hubert came back to me and told me he'd like to do the job himself. But if we were going to do something as risky as an Arctic crossing, we needed to do a test run first, to see if we could handle such a tough expedition. We chose Greenland.

Our intention was to walk across Greenland, diagonally, from the south-east corner to the northwest corner, a distance of about 2200 kilometers. All the way to Thule. Right at the start, we were caught in a series of terrible storms and were tentbound for six days in total. We came close to giving up and going back, just like I had done once before in the Arctic in winter. Instead, we went on. The weather stayed bad, but the wind was frequently good for sled sailing.

H: So you managed to perfect your sled-sailing technique?

M: Hubert is a fantastic skier, much better than me. I had more experience, but he soon made up for that and together we made a great sled-sailing team. I took the compass and sailed ahead, and Hubert followed me, making sure he never lost sight of me. As the rear man, he was in charge whenever visibility was poor. The better skier, or in this case the better sled sailor, always has to stay at the back in bad weather if you want to go fast. We developed a method that worked for us: sail until we were completely frozen. It was always windy, always cold, and generally foggy. We could only cope with the sailing for a few hours at a time.

H: What did you do to prevent hypothermia?

M: We walked for a few hours to warm up. Then it was back to sailing. Walk, sail, walk. On our feet for eighteen to twenty hours. Very strenuous. Sailing makes you cold, sled hauling warmed us up again. We made fast progress with this method, 150 kilometers a day or more, up to a maximum of 180 kilometers in a single day. We really covered some distance!

All in all, it went very smoothly, maybe even too smoothly. Hubert was in great form. It was a brilliant trip, and we both had a good time. Perhaps our success made us a little bit arrogant. But we had managed to cover a distance of 2200 kilometers in thirty-five days, after all.

H: On one occasion, the wind tore your sail off and blew it away, never to be seen again. Did you feel the success of the expedition was in jeopardy when that happened? What did you do?

M: When we lost the sail, we knew right away that the walk-sail-walk game

Ice walking: snow and fog in Antarctica (1989)

was over. But we still carried on. It was harder going but less dangerous. When you're sled sailing in the fog it's easy to lose sight of one another. We each had two sails, one large and one small. Using different sails would have meant us drifting apart, and the wind was too strong to use the large ones safely, so we sailed piggyback using one large sail. It was difficult for the man at the back, but it worked. Until we got into crevassed terrain. From there on, we mostly went on foot.

H: What did you mean when you said that the speed you did it in made you a little bit arrogant?

M: When you are testing the limits of what is possible, you are constantly living between setbacks and confirmation. In spite of the bad weather, we set an incredible pace, all in all. We made a diagonal crossing of Greenland in a faster time than most people had taken for the short route, even when they'd sled sailed part of it like we did. This success made us cocky. We were pretty sure we'd manage our next project, the 2000 kilometers from Siberia to Canada, without any major problems.

H: After your quick success on the trans-Greenland expedition, you were already planning your next adventure, the traverse of the North Pole?

M: Yes, in 1995 we wanted to go from Siberia over the pack ice to the North Pole and on to Canada. That's 2000 kilometers as the crow flies. Across the worst pack ice in the world. And there's the drift to contend with, too.

This traverse of the Arctic was one of the last great problems—a big challenge, in other words. It had never been done without air support. But that's exactly how we wanted to do it.

H: How thick was the ice up there?

M: Two meters in the middle, but at the edges, where we set off, it wasn't even 20 centimeters thick. It could only take the weight of one of us at a time. It wobbled when you walked on it, and we could hear the water underneath. The ice skin rose and fell; it was like a blanket made of ice. It's a strange feeling at first, walking on ice that thin, but you get used to it. You get the feeling the ice is alive.

H: How can ice that is 20 centimeters thick wobble?

M: It's not like the kind of ice we get. It's not freshwater ice; it's saltwater ice—compressed and soft, like a firm mass, made up of little pieces. It's not like the frozen surface you get on our ponds in winter.

There are pieces as big as a soccer pitch and pretty stable. Other parts are grey, new ice, only about 5 centimeters thick—big strips as wide as a road. In between are channels, wide cracks in the ice that you have to try to get across. Taking to the water is not really an option.

Our main problem was the ice compression, the pressure ridges. There was a constant north wind that made the ice, and the water under it, unsettled. That north wind slowly increased to storm force, and we were stupid enough to keep going in spite of this. Instead of going back and waiting for things to calm down, we just kept walking north, into the storm.

It was fifty degrees below zero and Arctic night. We were walking to our doom, quite simply. During the second night, the ice around us compressed like an accordion.

H: You can't swim, can you?

M: There was no need to. I had such a thick suit on that I wouldn't have sunk. There was enough buoyancy to keep me afloat. The problem is, when you get out, you're just a solid block of ice. And thawing out drains a lot of heat from your body.

H: Was there any danger from polar bears? Did you have a rifle with you?

M: Of course. We had to scare a polar bear away with the rifle right at the start of the trip. The big fellow was clever. He hid behind an ice ridge and watched us, holding his paw over his nose. Polar bears hide their noses because that black dot is the only thing that can give them away in the gleaming white surroundings. They're incredible, those animals. The bear kept circling around us so it could attack us from behind. We had trouble working out where he was.

H: Did you take a shot at him?

M: No, no, no, that's very dangerous. We fired the Winchester into the air and set off some signal flares.

H: And? Did that impress him?

M: He reared up on his hind legs. He must have been at least two and a half meters in height, maybe 400 to 500 kilos in weight. An impressive animal. Luckily, he was frightened of the flares.

H: How did you and Hubert get along?

M: Very well. Hubert and I have been on about ten expeditions together. He was never a summit man; he always came along as expedition doctor.

On the Arctic trip he was a full member of the team. That meant we were responsible for each other. Things only got dangerous a couple of times, once at the beginning of the trip and once near the end.

He'd really enjoyed our previous expeditions and was eager to come with me to the North Pole. I was against the idea at first, as I was worried about how our mother would cope, having already lost Günther on Nanga Parbat. But after the Greenland trip I thought Hubert and I would manage. We made a great team. And so we decided to risk it. Hubert took three months off work.

We started in Siberia. There was another expedition there, but they wanted to try it with dogsleds and a dozen depots. They set off on the same day as us but from a different place. They'd previously flown the trip in a helicopter, so they knew the best route to take, but in spite of that they turned back after just a few hours, whereas we in our arrogance battled on into the storm.

We were heading for disaster. The ice around us started to give way and break up, as a belt of ice 80 kilometers wide was compressed to form pressure ridges. The plates of ice rose up and slid over and under each other. Between them was open water. It was pure chaos.

H: You were taken by surprise. You were in your tent; it was ten to three in the morning, Arctic night.
M: The ice was creaking and cracking and crashing so loud we couldn't hear ourselves speak. It was like being in a metal factory.

H: Did Hubert show any sign of panicking?
M: Hubert took a quick look outside and said, "The ice is cracking everywhere." We got out of the tent immediately and shone a flashlight around, but we couldn't see very much. Then we noticed that the ice we were standing on was sinking slowly. We dragged the tent as it was to a thicker island of ice and went back for the sleds. It went on like that the whole night long: pitch the tent, move the tent, drag the sleds, over and over again. At one stage we lost one of the sleds, and Hubert fell in the water.

H: You must have thought you'd never get out alive.
M: We told each other that night that if we made it out alive we'd never

do anything like that again—in other words, no more pack ice! Then we turned back and headed for the coast.

As soon as the ice we were on was more or less safe, we sent a distress signal via a satellite link. We didn't have a radio connection or a satellite phone; we just had an Argos device, the same type of device used to tag whales and track their movements. It works via a French satellite system, and the data is sent to a processing center in Toulouse.

We'd programmed the fifteen buttons on the device with a kind of Morse code alphabet, which included a signal for polar bear danger. On the first evening, when we had our encounter with the polar bear, we sent a signal, and the news that the Messner brothers had been attacked or eaten by polar bears appeared in all the papers.

H: How did you get rescued?
M: The Russians who dropped us on the ice had a helicopter parked at the nearest Arctic research station, ready and waiting for such an emergency. This precautionary measure was part of my logistical planning. The helicopter was sent out, and they found us right away.

H: Where were you?
M: We were in the tent when we heard the helicopter.

H: You waited in your tent for a helicopter while outside all hell was breaking loose?
M: At fifty degrees below zero, you can't just stand around waiting. You have to do something. You can walk around or pull a sled, but if you stand around, you get so cold after half an hour or so that you just start shivering. Then everything starts to seize up, you can't get your mouth closed, and the cold starts to kill you. It's dangerous out there.

H: And Hubert fell in the water?
M: Yes, it all happened in that one night. I pulled him out. He was a solid block of ice when he got back to the tent.

H: What kind of suits were you wearing?
M: We had three layers on. The outer layer was a really thick down suit. It

was pretty much waterproof, although some water got in at the ankles and through the zip.

H: That was the first time you had ever called out a rescue team. Did it feel like a winning streak was coming to an end?

M: What do you mean by "rescue"? We called out our helicopter, and that was part of the logistics, more or less. The helicopter crew was instructed to wait at the last station on call until we reached the central pack ice. That had been agreed in advance with the Russians; it was in the expedition contract. We set off in the knowledge that the helicopter was at our disposal if required. I knew that the major problems would come at the start of the trip. As it turned out, we'd never have made it back on our own.

H: Nevertheless, your survival was dependent on the help of others. How did that affect your feeling of invincibility?

M: I lost my youthful feeling of invincibility after the Nanga Parbat tragedy in 1970. And after getting flown out, we swore we'd never set foot on the pack ice again. Four weeks later we walked to the Pole from a floating station in the middle of the Arctic Ocean. It was nothing special. However, I felt then that the time for pushing the limits in extreme situations was over.

In 2004 I was sixty—the age that other people take retirement—when I decided to walk across the Gobi Desert.

H: You had planned on taking two months for the Gobi trip.

M: I'd planned on six to eight weeks, but I was faster than I thought I'd be. It didn't all go smoothly, but it went better than expected. To begin with, I was able to ride some of the way—horses and camels—and the Mongolian border police helped me out toward the end.

On two occasions I reached the limits of my abilities: in the southern Gobi, after 300 kilometers of walking without finding any water, and at the end of the trek, when I had to walk over the Altai Mountains back into the heart of Mongolia. I was totally drained. I have now come to terms with the fact that I am getting old.

CHAPTER IV
LOOKING FOR
ANSWERS

1995–2009

*Mountaineers have always felt that they are
special. The aura of death and danger that surrounds
them has elevated them to the elite. They must attain this
elite status once more, not in death but in life. The realization that it is by using less technology, not more, that
the last great problems of mountaineering will be accomplished would make mountains out of climbing gyms
again, make cathedrals out of the mountains, and
mountaineers out of mountain tourists.*

—Ralf-Peter Märtin

THE WILDERNESS MYTH

In 1999 I ran for election to the European Parliament as an independent candidate on the Green Party list in Northern Italy. During the election campaign, the body of George Leigh Mallory reappeared. The hero of Mount Everest had been dead for seventy-five years, yet he still lived on in my mind. When the photos of the body appeared on the internet, I imagined the scenario of Mallory's summit attempt on June 8, 1924. The images in my head were so vivid it was almost as if I had been there myself.

The challenge for me now was to find a way of telling Mallory's story. An amateur German historian and geology student, Jochen Hemmleb, formulated a theory about where the body might be found and the events surrounding Mallory's summit attempt, but these were merely preconceived ideas until the proof, as he saw it, finally emerged. Although he did not find Mallory's body himself, Hemmleb deserves great credit, since it was the discovery of that bleached corpse on May 1, 1999, that allowed the Mallory myth to grow. Indeed, it is thanks to his perseverance—and a lucky break, when Conrad Anker came across the body in a place that Hemmleb had not asked his team to search—that Mallory now lives in our imagination once more.

The Mallory myth is all about the unclimbed Mount Everest, an untouched mountain wilderness that repulses the advances of human beings and remains inaccessible to them. [Sir Edmund] Hillary and Tenzing [Norgay] were the first to climb it, but it was not Mallory's mountain that they climbed.

In this late phase of my life, I found myself thinking more and more about another dimension of the wilderness experience. As I came to understand the relative meaning of success, I focused on values of a different kind— tranquility, vastness, inaccessibility—the kind experienced by Mallory. As a result, I also acquired a new set of heroes: George Leigh Mallory; Fridtjof Nansen, who failed to reach the North Pole; and Ernest Shackleton, whose trips all ended in failure but whose exemplary leadership allowed his expedition to make a successful retreat from the Antarctic wilderness, a story that is without doubt more exciting than all of the success stories of the other adventurers.

I knew that trips like these would only happen in the future if parts of the world remained untouched. I also knew that the wilderness was shrinking fast. The international organization Mountain Wilderness, which I helped to shape, was founded in 1987 in Biella, Italy, and is dedicated to the preservation of wilderness mountain areas and the principle that "wilderness" as a value in itself is absolutely essential.

The yeti story would not be possible without wilderness either. The story I wrote in 1997 is a legacy of the mountain tribes of the Himalayas, a self-woven legend, a report and a myth in which every detail symbolizes the character of a tribe. The time will come when the yeti story is seen as a symbol of our conflict with the wilderness.

The Himalayan yeti legend is several thousand years old and originally concerned a large humanlike creature—part primate, part bear—which represented a constant threat to the newly arrived mountain people and was believed to live somewhere out in the wilderness. The legend grew in the telling and was continually embellished until it became the story of the yeti. It was only because both elements of the story had survived—the legendary figure of local folklore and the snow bear in the wilderness—that I was able to compare them and prove that they were congruent.

When trying to answer the yeti question, it wasn't the simple solution to the puzzle that most surprised me but the willingness of many Europeans to dismiss the obvious and believe the improbable. It seemed that, for us Westerners, invented reality had become a more powerful force than the real wilderness—if that still exists.

H: You say that good thoughts come when you are out walking?
M: I see myself and the world more clearly; I feel at one with the world. Walking also clears my head.

H: What happens to you when you walk?
M: It cleanses my mind and my spirit.

H: How does that work?

M: I can't explain it exactly. Maybe it's just better blood supply to the brain. But first of all, I like walking; secondly, I like walking in the open countryside; and thirdly, I like long walks. Half an hour is not enough for me. When I was serving my term as a member of the European Parliament, this need for exercise was a problem. You can't just leave a meeting at the European Parliament, wander outside, and follow the rest of the debate on your phone. Going for two-hour walks was out of the question. I don't know how those politicians manage to clear their heads. I couldn't.

H: Where do you and your family live now?

M: We've had the same arrangement for the last ten years. During school term time we live in my wife's apartment in Meran and we spend the summer at Juval.

H: Your son seems to be more at home in your castle at Juval.

M: Yes, he's got his horse there, and he used to have camels and llamas as well but he's not that interested in them any more. He used to train animals and then he went on to study biology. He climbs in his free time. He had a bit of an accident and dislocated his shoulder—it was nothing serious though.

H: Has your life changed since having a family?

M: I already had a child, but I didn't have a family. I never lived with the mother of my first daughter. She studied in the United States, and she and her mother now live in Canada. We met in 1979 and did some traveling together. We were together for a few months. Her name is Nena, and our daughter's name is Láyla. They both lived in South Tyrol for a few months. But it was never going to be a long-term thing. That would have been impossible. We lived in two different worlds, but that wasn't because of the child. We didn't grow apart; we just didn't belong together, even though Nena was a born adventurer, like an American Indian.

H: Why didn't it work?

M: It was impossible. That's all I can say.

H: Why was it impossible?

M: She tried to change me. We both wanted to constrain and confine each other.

H: Where did you meet your current partner, Sabine Stehle?

M: At a book launch in a mountain hut in East Tyrol. For Sepp Mayerl's book. She happened to be there. We only saw each other briefly; it was only later that we really got to know each other.

H: It doesn't appear to be an easy relationship either.

M: It's a fantastic relationship. We are a strong couple.

H: A merger is out of the question then?

M: I think that's part of our success story. Just because we are a family doesn't mean that everything has to belong to everybody. My partner, Sabine, finds it better that we each have our own responsibilities. She has a very clear idea about how she wants her apartment to look and how best to organize the practical things in her life, like what time to put the kids to bed, for example.

H: So there is a very clear and ordered system?

M: One time I gave her a few pictures for her apartment. I hung them up for her, and she changed them around. I knew she would.

H: Do you feel attracted to her because of her independence?

M: I like her tastes, her individuality, her character. Our daughter Magdalena said, "Did you see that Mama changed the pictures around?" "Yes," I replied, "I saw that." "Don't you go changing them all back again," Magdalena said with a smile. We all know that Sabine has her own way of doing things.

H: You are a wealthy man. How can you prevent your children from acquiring the attitudes of the nouveau riche?

M: My children are not nouveau riche; their mother sees to that. And they also see how I live. They see that the only assets I try to realize are my ideas, that it's never about money or possessions. The children know that their father has nothing but ideas and that he tries to implement them with everything in his power, as long as he has sufficient means to do so.

The children do not take our lifestyle for granted; they are unpretentious, thrifty even. For example, Simon asked me whether our trip to the Sahara Desert wasn't too expensive. "Yes," I said, "it's expensive, but I can finance it by writing an article about it. And because we both want to do it, it's worth the investment." Simon wanted to go; he planned the trip, and I put up the money. "You've done the research," I told him, "so now we're going to do it. Because we can, simple as that." He got less for Christmas that year. He didn't want anything else apart from that trip.

H: Which values have you tried to instill in your children?
M: The children know that the castle and the farms will never make a profit in economic terms—we are happy if we don't make a loss, in fact. They are purely a labor of love, like everything we do in life. So the way I see it, the castle and the farms will have to be passed on to an enthusiast, whoever that might be.

Other than that, our children are very sociable and not at all self-important. They would never go around saying, "My dad's the king of the mountains," or anything like that. That wouldn't go over very well anyway. But I have Sabine to thank for the children's upbringing.

H: Why do you remain on the sidelines when it comes to bringing up the children?
M: That's the way it's always been. The women took care of the children; the men moved on. The women found themselves a man; the men didn't go looking for a woman.

H: Don't you think you're making things a little too convenient for yourself with that attitude?
M: I don't want to be the one to say what the children can and can't do at home. Sabine decides, and we abide by her decisions. The children actually tell me, "That's not for you to decide," and that's fine by me. They wouldn't know what to do if they had to cope with different parental directives. That's why Sabine always has the last word at home.

H: You say that you don't even know how to make yourself a cup of coffee. I have heard of men who are incapable of boiling an egg, but not being able to make coffee . . . ? Do you reject the idea of housework as a matter of principle?
M: No.

H: You manage to cook spaghetti in the desert, and on Mount Everest you spent three hours melting ice to make two mugs of tea, yet when you are at home you can't even make a cup of coffee. That seems rather peculiar, Herr Messner.
M: Well, I've had the good fortune, or maybe the misfortune, of a lifetime of being spoiled by women. My mother spoiled us; she made our breakfast, and when we came home there was always a meal waiting for us. Then there was Uschi, who was a very good cook.

But when it comes to family-management matters, Sabine is unsurpassed; she is also a very good cook. She makes her own jam and lots of other things, from syrup to patchwork quilts. Our children never want to go out to eat; they say it's much cozier and the food is much better at home.

H: That still doesn't explain why you can't make coffee.
M: I can't be bothered. When I'm on my own, I'd rather go out for breakfast. I don't do household chores. Not because I can't—I was quite a skilled craftsman as a youngster when we had our chicken farm—but I suppose I'm more talented as a mountaineer.

H: You support the theory that the mother ought to look after the children rather than a housekeeper?
M: We currently have a housekeeper, the best we've ever had. She's well integrated into the family. But the children's mother is still always there for them, and they know that. The children need their mother. A nanny can never be as close to the children as their mother. I do realize, of course, that these days most families can't afford a housekeeper, which is a shame.

H: Do you predict an impoverishment of the middle classes?
M: It used to annoy me sometimes when I sat on EU [European Union] committees listening to the fine words of my colleagues. I like Romano Prodi, but when he says that in ten years' time Europe will be the leading

economic power in the world, I think about what I have seen on my travels. Just look at what is happening in China and India—and in America. Any realist can see that Prodi's statement is false—unless miracles happen.

We Europeans have no chance of maintaining our current standard of living if we carry on as we are. Our labor costs are too high, we've become spoiled, and we haven't even come close to making the structural changes necessary to ensure that Europe remains competitive in the global economy. In the last ten years it has become much more difficult for the average family with two or three children to make ends meet.

My parents had almost nothing; they earned a little bit of money with the chickens, and my father had a tiny income. In spite of this, they were still able to bring up nine children. In today's Europe that would be unthinkable. To bring up nine children on an average income—and send all of them to college—would be impossible.

H: At first, you had big problems with the South Tyrol bureaucracy concerning Juval. Why did they not want to let you live there?
M: At first they tried to prohibit me from doing pretty much anything. They wouldn't let me put a roof on the ruined building, then they refused to let me build an access road, and finally I was told I had to remove the solar panels I'd installed on an outbuilding.

H: Why do you think you had all these difficulties?
M: I don't really know. A power struggle maybe? The old mayor of Kastelbell versus the new guy? Whatever the reason was, I was certainly restricted when I moved from Villnöss to Juval. They didn't want me to live there. They said I had to connect the castle to the sewage treatment plant in the valley first. Either that or wait years for the certificate of habitability. The stupid thing was, back then there was no sewage treatment plant in the village.

H: What did the authorities have against you?
M: I don't know. I had already installed a septic tank in Juval, a three-chamber system. But the old mayor wouldn't budge. So I decided to stand up for myself. I told him, "Let's make this really simple. I'm going to visit all the farms in your borough, and if I find any that don't have proper

sewage treatment facilities I'll report them so you can press charges." It was only then that he finally backed down.

Since then I've been known locally as a blackmailer. Every time I go into the village someone says, "There he is, come to blackmail us again." I've never blackmailed anyone; I just stand up for my rights.

I own one of the oldest houses in the area. And I have rebuilt and restored it as carefully as possible, in accordance with the current regulations. But it sits 500 meters above the valley, so I can't hook the place up to the local sewage treatment plant. Juval is built on rock, perched on top of a 500-meter-high cliff, so that wouldn't be technically possible. That's no reason to let it become a ruin, though, is it?

The mayor and the local councillors are pretty reasonable now, particularly the young mayor. We're on the same wavelength. He's a realist and free from prejudice. I no longer have any problems with Juval.

H: When did you decide to allow tourists to visit Juval?
M: When my daughter Magdalena started to go to school. It was obvious then that we'd be able to live at Juval for three months a year at the most. That meant the castle would be standing empty for nine months. So I redesigned the place as a museum. And it worked well right from the word go. There were a few growing pains, naturally. It took a while to find the right staff, for example, but Juval is now a full-fledged tourist attraction.

H: How much does it cost to get in?
M: I think it's eight euros for adults [in 2004]. The entrance fee is enough to pay for all the staff, the insurance, and the upkeep of the castle. Juval is still our summer home, and that costs a little more. As I said, it's a labor of love that could be passed on in its current form.

H: What would you have done if your partner, Sabine, had been against the idea of the castle in the first place? What would have been more important to you—Sabine or Juval?
M: I certainly wouldn't have given Juval up right away. But when it came to the design and renovation, we had a lot in common. Any other woman would probably have told herself not to get involved with a man like me and his crazy ideas. But if a couple wants to build a strong relationship, a

love story alone is not enough. They need to have other things in common: travel, responsibility, shared goals.

H: Do you and Sabine share any similarities?
M: Well we have the same star sign, Virgo. And we both like things neat and tidy.

Many people assume that I have no structure to my life, that I'm a semi-nomad living a random life on a wing and a prayer. Outsiders probably have no idea how difficult a seminomadic life can be. I have to have a firm structure to my life; otherwise, I wouldn't be able to survive financially.

Sabine understands that, and she covers my back. We both love antiques and art, as well, and we have similar tastes when it comes to literature. And there are the children, of course.

H: What actually happened the night you fell off the castle wall and broke your foot?
M: The owner of a local restaurant called me and said, "Hey, Günther Jauch is in South Tyrol, and he wants to meet you for dinner this evening. Are you free?" We'd just gotten back from a trip, but since we don't get to see Günther Jauch that often, and since he was on vacation in Naturns with his family, and his kids are the same age as ours, I thought we could all have a nice meal together. So I said we'd be there at seven.

It turned out that the guy at the restaurant had said the same thing to Günther Jauch. He'd phoned him and said, "Hey, Messner's coming for a meal at my place this evening." We were being set up because the owner of the restaurant needed the publicity, but we didn't know that at the time. We ate dinner together, drank a bottle of wine, and chatted, and around nine o'clock we called it a night. Jauch headed back to his hotel; we set off back to Juval.

It was early summer. It was cold, and we'd had a bit of rain during the evening. When we got home, the key to the main gate wasn't in its usual place. I thought one of the staff might have taken it inside, so I shouted for a while, then searched around for it. Nothing. Then I said to Sabine, "I'll climb in around the back and open up for you." I climbed up the enclosing wall on the west side, but it was wet and dark, and when I was climbing down the other side, I slipped and fell. I was probably being a bit too blasé

The myth of the yeti: a Khampa dances with the stuffed head of a Himalayan snow bear, or Tibetan brown bear, the zoological equivalent of the yeti.

about it, to be honest, as I'd climbed it plenty of times before, but only ever in daylight. I should have told myself to be careful, but the thought never even occurred to me.

The second I slipped off that wet little hold, I knew it would end badly, but by then it was too late. I couldn't see the landing, so I couldn't bend my knees to absorb the impact. The next thing I knew I was lying on the ground, unable to walk, in the dark and the rain. I couldn't stand up. I shuffled down to the courtyard on my backside, with my hands behind me and one foot on the ground, and screamed for help. Sabine also started shouting, and eventually we woke up a staff member, who had been asleep in the eastern tower. He unlocked the gate, and I was taken to the hospital by ambulance. I had lost two liters of blood. It was an open fracture of the right heel bone, the calcaneus. Part of the bone was sticking out. A multi-fragmentary fracture of the worst kind.

H: Did the doctors give you any hope that an operation on the heel would enable you to continue with your extreme expeditions?
M: Not at first, no. They immobilized the foot and told me I had to rest it. My brother Hubert came to see me and told me I could forget about expeditions like the one to the North Pole even if the best surgeons operated on me.

H: Why was your brother so pessimistic?
M: A calcaneal fracture is the worst thing that can happen to a walker. And my calcaneus was shattered, in a hundred pieces. None of the doctors dared to operate at first. I was going to go to Switzerland to have it done until I finally found a young doctor in Bozen, a Dr. Waldner, who agreed to attempt an operation. He took the X-rays into the computer room and established which parts of the bone he'd have to replace—to remodel, in other words. Then he sawed off a piece of my hip, shaped it, screwed it together with the rest of the bits, and reinserted it into my foot. It was an incredible bit of surgery, a remarkable achievement.

H: Was this accident, and the fact that you were unable to undertake any big expeditions for a while, the reason why you started to become preoccupied with the yeti story?
M: Yes. But the yeti story had begun to interest me ten years earlier, in 1986, during an attempt to trek across eastern Tibet. The idea was to retrace the route of the great Sherpa migration of the sixteenth century—and to have an adventure with Sabine. She deserves a knighthood for that trip. Afterward, her father complained about what I put the girl through.

H: Why the knighthood?
M: Well, first of all, we traveled without a permit—and in the middle of nowhere. Secondly, we were arrested and held in custody. Thirdly, they

took our passports off us, which meant we couldn't go anywhere; we were stuck for days in a police station with no passports. I had no idea how to get us out of the mess.

Then Sabine said, "Looks like we'll just have to go for broke. Go tell those officials you're going to take photos of them and complain to Beijing about them." They were Han Chinese, of course. So I went up to the desk. The policemen had our passports in front of him. I pulled the camera out of my rucksack—I made a big show of it—and started taking photos. He grabbed his jacket, covered his face with it, and took off, leaving our passports lying there. We picked them up and left right away.

There were a few moments like that on the trip. We were always running away or being pursued. We were in a region that was prohibited for Europeans, an area as big as Bavaria. In the end we hid in a monastery, and I decided to go on alone. Sabine would never have managed the trek from Tachen Gompa to Lhasa. It was too far, too dangerous, and too tough, up and down through the mountains and valleys. We decided she should head back to Chengdu on her own. She got sick, but she made it to Lhasa.

H: Why didn't you go back with her? After all, the situation could have been life-threatening.
M: Because I knew she'd make it back on her own, and the trip I wanted to do was still possible in spite of everything.

H: Was your desire to complete the trip greater than your protective instinct?
M: We'd hired a man in Kathmandu to act as our interpreter and guide. He had disappeared, but I was sure he'd be back.

H: Didn't she insist that you go back with her?
M: No, she was really brave. I told Sabine to stay put and wait for our guide to turn up and then to go to Lhasa and wait for me there. Unfortunately, the guide never showed up. She found herself another guide, a Khampa, who was almost 2 meters tall and so strong that he commanded everyone's respect.

On the way back they came across the bodies of two Americans who had died when the truck they were in went off the road and down a hillside.

They'd been planning to do a trip similar to mine, apparently, to trek across Tibet. Sabine felt sick when she saw the bodies and all the blood. In Lhasa, suffering from dysentery, she checked into a hotel. She'd lost so much weight she was just skin and bones.

H: So you left your wife behind and went on alone. How did your relationship manage to survive that?

M: I spent a day and a night looking for her in Lhasa, but I couldn't find her anywhere. I started thinking she must have left. I had my own passport, of course, but she had the money and the return airline tickets. Finally, I found her in a hotel, where they'd accidentally deregistered her. She hadn't left her room for days. She looked like a ghost, almost transparent. When I found her, I didn't give her a hug; I said, "Hey, you, I've just seen the yeti!" She still hasn't forgiven me for that.

That trip was an intense experience, but it created a lasting bond between us. The kids think it's good that we managed to get through it all together.

H: Where did you see the yeti-like creature?

M: I was tramping through the wilderness, high up in the Kham region. I was never quite sure of the right way to go, so I kept asking the locals for directions. I'd get to a little village and seek food and shelter from the locals for a couple of days. I paid them for it, of course. Often, I ate with them and bought food for the next stage. If a yak caravan happened to be passing, I joined it.

Anyway, this one time the caravan didn't arrive and the villagers sent me west, along a river and into the mountains. The river was in spate, a raging torrent. I managed to get across it a little farther upstream, but there was nothing on the other side, no track, not even a set of footprints. I wasn't brave enough to wade back across the river, so I kept going. After a few hours of walking, I came across some footprints, then nothing again. It was getting dark, and I didn't know where I was. I wasn't scared of getting attacked by wild animals or worried I might go hungry; I just wanted to know where I was. I was hoping some locals might come along and help me out.

All of a sudden, I saw a movement through the bushes in front of me and something dark on the ground. "Yaks," I thought, "and if there are yaks, there must be people, too." The yaks always walk ahead; the Tibetans walk behind them, whistling. "And if there are locals coming, I'll be okay," I told myself. But there was no sound of bells, no shouting, nothing. It was puzzling. You always hear people when yaks are around.

Then, suddenly, a huge black creature appeared out of the dark forest. It walked out of the undergrowth, then it was gone again. It simply vanished without a trace. It was too dark to see it clearly, but I registered that it was big and black. I stopped and listened, sniffed the air and looked around. But the second the creature appeared, it was gone again. I couldn't say what it was or how big it was, just that it was bigger than me, black and shaggy.

For a moment I felt scared, but after a second or two I went over to where I'd seen the thing. The ground was clay, and I could see footprints, like human footprints, only much bigger. My first thought was that they looked exactly like yeti footprints. I took some photos, but I didn't want to spend the night there, as I knew the yeti must have seen me and he might come back. It must have been around midnight by then. I set off walking and eventually got out of the forest and onto a high plateau.

Suddenly another creature appeared in the moonlight. It was standing on two legs. It was quite a distance away, but it still looked frighteningly large. It might have been the same one I'd seen before that had trotted on ahead of me. I was so scared that if it had come toward me, I doubt I'd have been able to move a muscle. Then it vanished among the jumble of rocks on the plateau. I was dead tired and wide awake at the same time. It was terrifying.

H: Did you panic?
M: I stayed relatively calm and kept shining my flashlight around. It was my way of saying "There is light here, fire." For a wild animal, a flashlight means the same thing as fire, and the creature I'd seen could only be an animal of some kind, albeit a big, powerful one.

Later that night I came to a stream I couldn't cross. I walked downstream to where it flowed into a river. I still couldn't get across, so I tried to build myself a bivy shelter out of rocks in the fork between two streams. I rolled out my mat, threw my sleeping bag down on top of it, and said to myself, "I'm stopping here." I'd been talking to myself out loud like that for hours. But I couldn't settle. The first noise I heard had me sitting bolt upright. I was so scared that I got up and packed everything away again. I was afraid that the creature might attack while I was asleep and kill me. I had no idea what it was.

I walked back upstream and found a bridge. Then, around two or three in the morning, I came across a village. I shouted—no reply. A pack of dogs came at me out of the darkness; they were everywhere. I grabbed a big stick and tried desperately to fend them off. Finally, I took refuge in one of the houses. I climbed up a ladder to the second floor, crawled into my sleeping bag, and fell asleep.

I woke up with a start when rocks started landing on me. I crawled out of my hiding place, switched my flashlight on, and saw a mob of locals standing there—twenty men at least, with flashlights, knives, and clubs. They started shouting, and I immediately realized I'd better do what they wanted. If I hadn't reacted the way I did—standing there in my underpants, shaking with fear and holding my sleeping bag and rucksack in one hand—that would have been it. I knew if I didn't go down, they'd kill me for sure.

H: Maybe they thought that you were the yeti?

M: Yes, exactly. They probably thought one of the monsters had come into their village, as had happened many times before, so they'd all flocked together and armed themselves in defense. I had no weapon, only my experience. I instinctively knew that I had to stay calm and comply with their demands. Otherwise, they would kill me.

H: How did you manage to make yourself understood?

M: The first thing I did was greet them—in Tibetan, "*Tashi delek.*" Then I tried to explain, using signs and gestures, that I'd had to run away from a monster myself.

That night was the first time anyone had ever told me that there was a yeti-like creature in that region, and that the local people called the beast "Chemo." I learned that the creature was just as large as the yeti portrayed in the legends and that "shaggy," "smelly," and "humanlike footprints" were commonly used to describe its traits.

When I got back to Kathmandu, I ended up interviewing dozens of exiled Tibetans, refugees with an intimate knowledge of Tibetan language, culture, and mythology. I asked them all the same question: "What do the Tibetans call the creature that the tourists and the Sherpas refer to as the yeti?" The answer was always the same—"Chemo." So I knew then that the Chemo and the yeti were identical. And it was a Chemo that I'd seen— there was no doubt about it.

I decided to do some more research, and in 1988 I returned to the same region of eastern Tibet. I stayed there for a month, but I never saw the creature again. The local people told me that if a Chemo sees a human waiting for it, it will remain in its hiding place. That certainly seemed to be true. In 1995, after I fell off the castle wall, I decided to keep on researching the yeti myth for as long as it took to find out what lay behind it.

I went on a dozen expeditions—as an invalid at first—and spent a great deal of money. I traveled to Tibet, Kham, Amdo, Bhutan, the Altai Mountains in Mongolia, the Gobi Desert, Nepal, and Kashmir, trying to track down the yeti. Everywhere I went, I asked the locals what they called the monster, and although I ended up with a list of around a hundred different dialect names for it, it was always the same creature, the same story.

H: Why did you go to all that trouble and expense when you could just have gone to Lhasa Zoo and seen a Tibetan brown bear?
M: That's right; I found that out later. And when the Dalai Lama visited South Tyrol, after I'd solved the puzzle, he told me, "I think we've got yetis in the zoo in Lhasa. I think the yeti and the Chemo are the same thing." I told His Holiness that I'd come to the same conclusion, that the yeti and the Chemo were one and the same, and asked him to keep quiet about it until I published my findings. He told me he'd always suspected it and that he'd never really understood what Europeans imagined the yeti to be.

H: With all due respect, your revolutionary scientific discovery has a limited impact when one considers that all you really managed to establish was the fact that the two names refer to a bear.
M: That's true. But you wouldn't believe the preconceptions I encountered. Identifying the yeti as a creature halfway between an ape and a human was basically the mistake of the industrial society, a typically European idea, an image created by screwed-up civilization fetishism. Nobody had ever taken the trouble to ask the local people how they imagined their yeti to be.

This story, a legend that was thousands of years old, with all the characteristics of that strange and terrible creature, is only possible where wilderness exists. Snow bears behave in many respects like humans: they walk on two legs, with their feet flat on the ground, and they eat what humans eat. But unlike us, they are nocturnal animals.

Essentially, all I did was establish that the yeti legend was derived from observing an animal, and that it was the zoological yeti that gave rise to the legend, whether it's called the Chemo, Dremo, or Miti. And the animal really exists—still. Its behavior, its size, its appearance, and its eating habits correspond exactly to the creature the local people call the yeti.

Eventually, some Khampas told me that yeti relics were once displayed in monasteries. The question was: were they real?

H: The Chinese destroyed almost all of the monasteries in Tibet, didn't they?
M: Most of them, yes. I think about six thousand were destroyed. However, I managed to find an old monastery that was still intact, high up in the mountains of Kham. It was a dangerous trip, and I was pursued by the Chinese, but I managed to sneak in and get to the monastery. And what did I see hanging by the entrance? A stuffed Chemo! I photographed it from all sides. It was a huge beast, more than two and a half meters tall, and it must have weighed around 300 kilos when it was alive.

H: After the yeti, you devoted yourself with great passion to the story of the mountaineer George Leigh Mallory, who died on Everest in 1924. Why was that?
M: After they found his body, I knew it was the story for me.

H: Why was that? The man had been dead for seventy-five years and you had never met him. Why did you suddenly feel so drawn to him?
M: For a start, because Mallory was the first climber to attempt Everest—he was a pioneer, a legend. And also because my mother had read his story to me when I was a child, up on the Gschmagenhart Alm, by the light of an oil lamp. It was my first mountain story, and therefore the most important.

H: How old were you?
M: Five or six. When I got home after my first trip to Everest, I said to my mother, "If I remember correctly, you read the story of Mallory to us when we were children." "Yes," she said, and a few days later she brought me

the little booklet she'd read to me from. We couldn't afford proper books back then.

Mallory was a thinker—he moved in the same circles as Virginia Woolf—an intellectual from high society. But he also had this crazy dream of climbing Mount Everest. When asked by journalists why he wanted to climb the mountain, the answer he gave them became famous. "Because it's there," he said. It was a bit cheeky of him, really, in an age that was determined more by necessity than by frivolity. I wouldn't put it that way myself these days, but back then, it was a good reply. And he was a fabulous writer. In the last fifty years no one has written as well about mountaineering as he did.

Mallory was actually born in the same year as Paul Preuss, my all-time favorite climber. He was a writer, too. They were both born in 1886.

H: How would you sum up Mallory's attitude?
M: Athletic, bold, cheeky, and with plenty of English understatement. English climbers have this way of downplaying heroic deeds. Playing the hero is dismissed as laughable, because they know that on the big mountains you soon get brought down to size.

Anyway, the Mallory story was planted in my subconscious mind when I was a child, and later, during my solo ascent of Everest, he became a kind of companion to me. I could imagine how he had lost his life, but there was more to it than that—I felt like I knew him. I actually said where I thought the body would be, and that's where they finally found it.

H: If you knew where he was, why didn't you go and look for him yourself?
M: I didn't need Mallory's body to write about his spirit. And for me, that's what it was all about—his spirit, his vision, his style. Mallory was found by accident, only because an inquisitive American climber by the name of Conrad Anker went off on his own, outside the designated search area. The others tried to call him back, but he kept on going and stumbled across Mallory's corpse.

H: Why are you so certain—contrary to the people who found the body—that Mallory never made it to the summit of Everest?
M: The man who found the body agrees with my conclusions. Anker is

enough of a climber to understand the context. From a mountaineering point of view, it can be verified that Mallory was not on the summit.

H: Because in your opinion the Second Step would have been impossible to climb in those days?

M: Yes, and in nailed boots, for sure.

In 1925 the best rock climber in the world was a man called Emil Solleder. He was the first to climb grade VI. With double ropes, pitons, and special climbing shoes. But that was in the Dolomites, at an altitude of 3000 meters.

Mallory didn't have any pitons with him, he was wearing the wrong type of boots for hard rock climbing, and he only had a thin rope that would have snapped if he'd fallen more than a few meters. So, grade VI rock at 8600 meters in 1924? Unthinkable. No one could have climbed that pitch back then. And Mallory was a clever man. He'd seen the Second Step from below, so he must have known you couldn't just climb it like you would climb a route on a crag in England.

H: What do you think happened?

M: Mallory and [Andrew] Irvine probably got as far as the Second Step and decided they weren't even going to attempt it.

H: What about the length of rope and the wooden tent peg that the Chinese expedition found on the Second Step in the 1970s?

M: That's just a rumor. The Chinese found some gear below the Second Step, not on the Second Step or above it. They also implied that they'd found some old oxygen bottles above the Second Step, which, if it had been true, would have proved that Mallory had gotten to the top and substantiated the Chinese claim that their 1960 expedition team had also made it to the summit. However, the oxygen bottles had actually turned up a long way below the First Step.

H: What possible interest could the Chinese have in crediting Mallory with the first ascent?

M: The Chinese were eager to prove their disputed claim that their 1960 expedition had climbed the mountain. By maintaining that they'd found some equipment used by the 1924 British expedition, they thought they had supplied the proof needed to substantiate that claim. But where is the evidence?

H: So you dispute the Chinese claim that they went to the summit in 1960?
M: The way they described the summit assault—in the middle of the night, with no flashlights and no oxygen equipment—it is impossible to believe. They also wrote that one member of the team climbed the Second Step barefoot and that they stood on each other's shoulders at one point.

What I believe is that the 1960 Chinese team realized it was impossible to free-climb that pitch. Why else would they have used ladders on it—aluminum ladders that they carried up in sections and then reassembled—when they went back, in 1975? But it was what Mallory did on Everest that I was interested in, not the Chinese.

H: You say that you see Mallory as a kind of soul mate.
M: That's right, there's a spiritual kinship. I only have to read what he wrote.

H: Why is that?
M: I don't know. It's just a feeling.

H: In your book you have Mallory writing in the first person.
M: Yes, it's a literary device. By the way, some of the sections that were attributed to me are actually direct quotes from Mallory's diaries and letters to his wife.

H: Yes, but you must have felt a certain kinship with the man. Otherwise, his first-person narrative is not a literary device; it's just you making assumptions.
M: As I mentioned, I quoted him word for word in some sections of the book.

H: In the book you imagine Mallory lying there, dead, complaining that his rest has been disturbed by sensation-seeking individuals intent on finding his body.
M: Yes, then I resurrect him and imagine him observing how expedition mountaineering has changed over the last seventy-five years. I like Mallory, his clear sense of purpose and his naïveté.

H: What made you so sure of the manner in which Mallory and Irvine met their deaths?

M: Mallory and Irvine failed at the Second Step, or maybe lower. Mallory said himself that the monsoon might put an end to their attempt, and they disappeared as the monsoon clouds moved in.

I think they were on their way down—maybe in the dark, maybe not—and that somewhere below the First Step they went off course. One of them slipped and pulled the other off with him. The rope caught on a spike of rock, snapped, and Mallory was left lying there, with a shattered foot, unable to continue. The scenario is certainly plausible. Things like that often happen on the kind of terrain you find below the northeast ridge of Qomolangma.*

And as for Irvine? Well, we still don't know.

H: Isn't your perspective of events just pure conjecture?

M: My version of events is certainly closer to the reality than Jochen Hemmleb's, who appeared on the scene as a historian and pursued the theory first put forward by the American historian Tom Holzel twenty-five years ago. Since then, Holzel has reexamined and rejected many of his original preconceptions. Hemmleb, on the other hand, describes scenarios that would appear problematic to any high-altitude mountaineer. Hemmleb simply does not know what it means to climb a vertical piece of rock 30 meters high at that altitude.

It is no wonder that climbing history is losing its power to inspire, when authors who have no experience of climbing take a stab at writing about it. The history of climbing is exciting, but it can only be written by climbers, good climbers. People like Walter Schmidkunz, for example.

* Qomolangma is the Chinese name for "Everest."

CHAPTER V
SHAPING THE
HOMELAND

1991–2004

*Politics is not static. It must be taken apart,
examined and only the good parts retained.
Politics is a process of constant change.*

—Alexander Langer

SOUTH TYROL IN EUROPE

It had never been my aim to get involved in politics in any official capacity, although I had been politically aware since I was a young man, asking questions and joining in discussions on local issues. Years later, together with my friend Alexander Langer, I developed a vision for a different kind of South Tyrol. Langer led the political opposition in South Tyrol and was elected to the European Parliament in 1989.

I spent the autumn of 1991 walking and climbing in South Tyrol, the place where I was born and grew up, in the company of mountain guides, friends, and artists. This "Around South Tyrol" tour was conceived as a kind of political initiative, a positioning exercise to examine who we South Tyroleans were and where we wished to go in the future.

In 1999 I ran as a candidate for the European elections. I had no party affiliations, but I campaigned in Northern Italy on the Green list, winning a seat in the European Parliament by the skin of my teeth. As a newcomer in the Greens/European Free Alliance (EFA) faction, my first experience of the European Parliament was as a backbencher. I was a Member of the European Parliament (MEP) for five years, from 1999 to 2004.

All in all, it was an important time for me, but as a creative person who refuses to give up my own vision of the future and my right to determine my own life, five years was enough. I haven't given up my Green way of life, only my seat in Parliament. It's not a case of breaking away from the Green Party, either—I was never a member in the first place, and I won't be joining any other political party. I am still concerned about environmental issues and am anxious to see more ecological justice. For me, the key issues are still sustainability, open spaces, and the right of the individual to lead a self-determined life. The focus of my political work is quality of life.

The degree of autonomy we have in South Tyrol—and the high level of subsidies, which are distributed by the governing People's Party—has not only brought about an increase in wealth and regional pride but an increase in envy, hardship, and dependency as well. When the majority fails to acknowledge the despair and vulnerability felt by the many minority groups in the region, it can lead to fear and to discontent, the kind of discontent that can easily spill over into aggression, and this worries me.

In general, small autonomous regions tend to implode when the political power structures are centralized like they are in South Tyrol, with its single-party government, media monopoly, and system of subsidy distribution. It is only when political responsibilities are matched by a growth in democratization, the transparency that comes with media diversity, and an improved system of wealth distribution that a region like South Tyrol will blossom and flourish—to the advantage of surrounding regions.

Many of the reformers have now gone, however. Some have emigrated; others have fallen silent or retreated into their own private worlds. Maybe the self-determining South Tyrolean no longer exists. It might be that the final answer to all these questions lies outside my sphere of influence and responsibility. However, going away or staying away is not the right solution to our problems, as that means absenting yourself from the passion for this region and its future, and ultimately failing to accept our shared responsibility.

The landscape of South Tyrol is one of the most beautiful on Earth. Where the orchards end, the forests begin, with woodland glades, solitary farms, and a peace and tranquility that has remained unchanged for centuries. Above them are the barren hilltops, weatherworn mountains, and nameless ravines. It is a beautiful land. Those who experience the region only on their vacations will find relaxation and enjoyment. They will be well looked after, too—we South Tyroleans are good hosts. At the same time, the apparent peacefulness of this land conceals a desperate plea from the villages for help, enlightenment, and knowledge. Yet nothing happens. The region is compact and manageable, for sure, but the manipulation of the people is huge and their lack of voice immeasurable.

Those who think there is nothing to lose that has not already been lost, who retreat into self-imposed exile, have not won; they have simply given up.

H: How did you come to get involved in politics?

M: The man who had a decisive influence on me was Alexander Langer. He was an alternative South Tyrol politician, an activist, and an idealist, who founded several political movements and campaigned in opposition to the People's Party of South Tyrol, who held a monopoly of power in the region.

H: Would you describe yourself as being left-wing before you met Langer?
M: I am a liberal person. But I have even less use for leftist ideology than I have for middle-class values. On many issues—landscape management, protection of the mountains, consumer protection—my opinions are actually conservative.

H: What do you mean by "leftist"—that the state takes care of everything?
M: To begin with, Langer ran around with Mao's *Little Red Book* even though he came from a middle-class family. He was involved in the '68 protest movement but quickly outgrew it. Langer was a contemporary of the radical left-wing activists [Joschka] Fischer and [Daniel] Cohn-Bendit. He maintained a close friendship with Cohn-Bendit and accomplished similar things in Italy as Cohn-Bendit did in France and Germany.

H: What exactly do you mean by the term "leftist ideology"?
M: "Real socialism" already existed in East Germany, the USSR, and China. In Italy the whole thing was less clear-cut, as we had a strong left in the form of the Communist Party, which at one time was the second-largest party in Italy and whose postwar leader was an admirer of Stalin.

Like social democracy, this Italian communism was a kind of socialist movement. The Italian '68 movement idolized Mao, but during my travels, and particularly on my later trips through China, I had become anticommunist. That's about as far as it went for me when it came to left-wing politics.

H: So the Chinese model was not a source of enlightenment for you because you had seen the reality?
M: I had experienced the realities of real socialism firsthand—the camps in the north of China, the derelict buildings, the dreariness. I had also registered what the East German newspapers wrote about me. I was regarded as asocial, reactionary—the worst of the worst, in other words.

Climbing mountains on my own ran contrary to the principles of collectivism. I was roundly condemned by *Pravda* for having climbed Mount Everest alone. I was and still am proud of that.

H: That surprised you? After all, the greatest enemy of communism is the individual.

M: Climbing a mountain by communist rules means you can only climb as a team, a collective. On the 1975 Chinese Everest expedition, there were about six hundred people on the mountain, a big, long convoy.

Under communism, autonomy was a sin. I had some lively arguments with Alexander Langer about things like community, the individual, property, and responsibility. As an individualist, I viewed East Germany with great suspicion.

H: Were you ever in East Germany?

M: Yes. I was invited over by some East German climbers in 1980. It was a huge amount of fuss just for a few lectures. There was a conspiratorial meeting on the Berlin autobahn somewhere.

They showed me some newspaper articles that had been written about me, the individualist, the egoist who didn't climb with the team but went off on his own and refused to obey the expedition leader. It was nasty stuff. Basically, it was the same kind of thing that Herrligkoffer had been writing but from a different angle.

H: How did you know when and who to meet? Did you have a secret set of instructions?

M: I don't remember. I do remember driving over in a VW Beetle, though.

H: Were you not scared of getting caught by the Stasi?

M: No, I wasn't worried. I was more concerned about not driving too fast on those terrible roads. Anyway, I found the people I was supposed to meet, and they told me that even though the newspapers wrote negative things about me, all the climbers over there admired me. They invited me to do a few lectures, but it turned out they couldn't get a venue because I was persona non grata.

I had more success after the Wall came down, and I traveled around Eastern Europe quite a bit. I knew a bit about the Eastern bloc climbing scene from the climbers I'd met on expeditions—Saxons, Russians, Czechs, and Poles—who I kept bumping into in Kathmandu or at various base camps. They were great people. They had a different approach than us, though. They got money from the state, and tinned food from the Five-Year Plan. Their food was better than ours, but their equipment wasn't as good. Their sleeping bags were all homemade. They were all good climbers, modest and unassuming.

H: So what did the Eastern bloc mountain food consist of?
M: They really did have the best of the best. The Russians, for example—they took caviar with them on the eight-thousanders. And they had oxygen bottles and ice screws made from titanium.

H: Does caviar make good mountain food then?
M: For sure.

H: Why is that?
M: You don't want to eat boring, tasteless food up there. That's no good at all. It has to taste good or it just gets left.

H: So the food needs to have a kick to it?
M: Yes. Otherwise, you can't be bothered to eat.

H: So the Russians had caviar, and you sat there with your ham sandwiches?
M: It wasn't so bad. We had plenty of food. We sometimes used to share it with the Poles. They were the first foreign climbers I got to know on my expeditions. Great people, and good climbers. I learned a lot about politics from them.

But it was Langer who first got me really involved in political issues. He was elected to the European Parliament in 1989. Prior to that he'd been persona non grata in South Tyrol. They'd treated him even worse than me, like the Devil incarnate, when he was actually a very caring man in the Christian sense. He was so open, educated, and enthusiastic—an incredibly nice man.

Even my children loved Alexander Langer. He didn't have any children of his own. Every time he was in the neighborhood, we invited him over for dinner.

When he became an MEP he moved away from South Tyrol, but he continued his leadership of the Green movement in Brussels, Strasbourg, and Bozen. Langer came from a Jewish family, and that was also used against him by certain people in South Tyrol. Even as late as the 1970s, he was referred to by some as "Langer the Jew," yet he remained an idealist in the truest sense of the word.

In 1994 or 1995 Langer decided to run for mayor in his hometown of Bozen. His dream was to create an interethnic town, where peaceful coexistence between the Ladin-, Italian-, and German-speaking groups was truly fostered. It was an important experiment with worldwide implications. Langer's primary concern was to bring together the German and Italian ethnic groups, and to create a new identity for the town. It would be a masterstroke of interethnic politics. But his candidacy was blocked, as he was unable to produce the certificate of linguistic identity required by law. In the 1991 census, Langer had refused to declare himself as belonging to any regional linguistic group, as he believed that this was counterproductive.

In 1994, before the European elections, he came to see me and said, "You should run for office as an MEP. You'll be elected. You'll get more votes than the other Green competitors on the list, so you'll win the seat. I'm going to go back into regional politics in Bozen, maybe even run for mayor." I told him it was too soon for me and that I didn't think he'd get the post of mayor, not this time anyway, although I could well imagine it happening at some stage in the future. As it turned out, they didn't even allow him to run. He was distraught. Finally, in 1995, he killed himself.

H: Did you have any idea that he might do that?
M: He'd dropped a couple of hints, but he never said anything directly. None of us could have anticipated that he'd take his own life.

H: Why do you think he did it?
M: He was unhappy. He'd experienced disappointments in his political and private life. Shortly before his death, he'd come into an inheritance, from his mother. She came from a rich, aristocratic South Tyrol family. She lived in a big house on the outskirts of Bozen. I was invited over for a meal once,

and Langer's mother voiced her concerns. "Herr Messner," she said, "I have invited you here because you are a practical man. My Alex has no idea about money. Perhaps you could talk to him." She had three sons. One of them ran a pharmacy, one was a doctor, and the third, the eldest, was Alex. She went on to say that all she wanted was for Alex to take his inheritance.

But he wouldn't. He refused to take even part of her estate. There was an orchard on the edge of town, about one hectare in size, which Alex's mother, with her sense of fairness, wanted him to have. I agreed with her. "Alex," I said, "it won't cost you anything. Take the orchard and lease it to someone." He didn't want the land, but he ended up inheriting it anyway. I don't think he really appreciated what was involved.

Anyway, the orchard then became zoned for commercial use. From one day to the next, it became worth several million euros. Every newspaper in South Tyrol ran a piece about "Alexander Langer the millionaire." The poor student leader was suddenly well-off, a rich man in fact. It was a blow for him. He had always stood up for the poor and the disadvantaged and campaigned for social equality.

H: It sounds like a first-world problem.

M: Langer called me. He was frantic. "Reinhold," he said, "I don't know what to do. They're doing a real hatchet job on me. You know I never wanted the land in the first place." I congratulated him and added, with a touch of irony, "Now we are both capitalists." He didn't find that funny.

H: Were you able to help him?

M: He was deeply unhappy. After he died, they set up a foundation in his name. I advised them to put at least half of the money from the sale of the land into the foundation. I can't remember if that actually happened.

H: It seems bizarre that anyone would kill themself because of a large inheritance and political problems.

M: Alexander Langer always gave everything he'd got. He always wanted to achieve more than was realistically possible. In human rights issues in Kosovo, for example, he didn't achieve what he wanted. The whole political process moved too slowly for him, and he thought the politicians were too timid to push through the changes he wanted to see.

He was a true idealist and a Christian, a sensitive man who could sympathize with the plight of others. Perhaps he was overloaded by the suffering he had seen. He only left one note, his legacy to the Green movement: "Keep doing the good things."

H: How old was he?

M: He wasn't even fifty. He hanged himself from an apricot tree, somewhere near Florence. On the anniversary of his father's death. For me, he was the most important postwar politician in—and for—South Tyrol. More important than [Silvius] Magnago.

H: The manner of his suicide seems very stylized.

M: I always feared that his idealism would prove to be his downfall. The world is not the way he imagined it to be, or the way he hoped to make it. But the world is not all bad either. In politics, as in life, the trick is to recognize and come to terms with reality and to change things step-by-step. Politics is all about persuasion and compromise. To insist that the world has to be a certain way is dangerous idealism. You can't force your ideals on the world; you can only work toward them.

H: Would you describe yourself as an idealist?

M: I try to be a realist these days. In my first book, there were a lot of idealistic images, statements that I wouldn't necessarily make nowadays.

H: What kind of idealistic statements?

M: The mountain world is clean and pristine, climbers are better people, that kind of thing. People who subscribe to clichés like that are ultimately members of a sect that, in the name of comradeship, seeks to ostracize those climbers who have a different view on life.

The typical members of the alpine clubs, often a kitschy value-based society, have become my worst opponents, because I point out these contradictions to them and remind them that this kind of polarization—this is good, that is bad; here is light, there are shadows—has all too often led to catastrophe.

I do not wish to have any part in this "climbers are better people" ideology. I would rather be ostracized than assimilated! I read recently that my greatest accomplishment was my unerring ability to make myself unpopular.

H: And you thought that was good, right?
M: Yes, I thought it was very good. It's also true. I never set out to make myself unpopular; I just say what I think, so I sometimes rub people up the wrong way, particularly people who parade their ideals before them like a banner and use them to hide behind.

And I will keep on saying what I think. I am not prepared to abandon my self-determined existence for an ideology or an ideal. I am a stubborn South Tyrolean, in that respect at least.

H: You always seem to be getting caught up in one quarrel or another.
M: I'm not prepared to play the whore for a few votes or applause. The journalists might be able to determine whether I get elected or not or how popular I am but not what I have to say. I want to be liked for what I do, not for what I pretend to be. I try—maybe too forcefully at times— to have the courage of my convictions and to stand my ground when necessary. That doesn't always go over well. But there are more than enough people who simply toe the line.

H: Surely the character profile you have just sketched is totally unsuitable for a career in politics?
M: Maybe. But it still won me my seat in 1999. Just six weeks before the elections, the Italian Greens came and told me that Alexander Langer had suggested I might run for Parliament. At first, I turned them down. Then I discussed it all with Sabine, and I came to the conclusion that it would be better to wait a few more years.

H: And what was Sabine's position?
M: She thought I should do it.

H: Why was that?
M: She wanted to see me in a position of responsibility. It's easy to talk politics when you're sitting at the kitchen table, but actually taking on that

responsibility is a different matter. Being an MEP means working for the common good.

H: Did money play any part in your decision?
M: No, money is just a means to an end for us. It provides us with the creative space we need. I told Sabine that if I became an MEP, we'd probably have less income than before. We are quite old-fashioned when it comes to our family life, so important decisions like that are always made together.

H: The MEP job would obviously mean fewer lecture tours and more time sitting around.
M: And canceling all my sponsorship contracts. But the challenge did appeal to me. Not because the other South Tyrolean with a seat in the European Parliament was my greatest opponent in my own country but because I believe we need to see Europe as a single community and integrate South Tyrol into this Europe. Not the other way around, emphasizing issues that are of particular interest to South Tyrol, as my colleague does—or pretends to do.

H: You mean Michael Ebner from the powerful publishing dynasty, whose family also owns the conservative newspaper *Die Dolomiten*?
M: Correct. I shouldn't complain anymore, though. The people of South Tyrol are now reacting to the paternalism and rebelling against a clan that only pursues its own interests and has crippled the region. With their combined interests in politics, the media, and the economy, the Ebners have bossed South Tyrol around for fifty years. That's long enough.

H: How did Sabine finally manage to persuade you to go into politics?
M: She asked me if I really knew what I wanted. Then, just to provoke me, she said, "You don't want to run because you're scared of losing!" With the Greens being as weak as they were in Italy, there was certainly a danger of that happening.

In spite of that, I had to do it. At the start of the campaign, I didn't have a clue about the political gamesmanship and the infighting that went on. Specific policy measures are one thing; trading political posts and power games are something else entirely.

Messner makes his final speech to the European Parliament (2004).

H: So all of a sudden you were forced to campaign, to make speeches in village squares in front of twenty people?

M: Exactly. And I wasn't used to that kind of thing at all. I'd given lectures in the same villages a year or so before in front of five hundred people, and now, as an MEP candidate, only twenty turned up. They probably wondered why I was even bothering to go into politics. You only realize how little politics matters to people when you are out campaigning. After you are elected, you get more respect—an interesting experience.

I used to ask myself why so many people want to go into politics. Maybe because it's lucrative. And it sounds pretty good, too: "Member of the European Parliament." At the time the only "outsiders" in South Tyrol who had managed it were Langer and me. And for me, one term was enough.

I still get ten letters a day from people who previously only ever voted for the People's Party, asking me for help. But in the majority of cases, my hands are tied—the provincial president is the person they should really be talking to.

H: What were your first impressions of the European Parliament in Strasbourg?
M: On my first visit I was impressed and confused at the same time. The Parliament building itself is a work of art from the outside, but inside it's confusing and hectic! Without my assistant, Max Rizzo, I would never have found my way around. Our office was a tiny little cell; it actually felt like being in jail at first. But I had no one to answer to, apart from my own conscience.

H: How would you describe your political mission as an MEP?
M: Well, first of all, I had to accept the fact that I was just a backbencher among 626 MEPs and that the European Parliament does not have a great deal of decision-making authority.

However, since I was more concerned about the bigger picture rather than regional and specific-interest issues, I tried not to get bogged down in paragraphs and procedures, and instead tried to raise concrete questions about a shared European consciousness, cultural diversity, and fostering peacekeeping initiatives. I also tried not to lose sight of my life outside the political arena. In fact, I was able to promote this vision of Europe outside the European Parliament through lecture tours and trips to Iraq, Kashmir, Nepal, and Cambodia.

I had hoped I might be able to play a part in inspiring a new era in ecological and regional politics, but to my disappointment that did not happen.

Nevertheless, my five years in Strasbourg and Brussels were an important experience for me. Cured of the notion that politicians can change the world, and with great respect for the pragmatists among them, I now try to influence things in a nonparliamentary way.

H: What did you manage to achieve in your time as an MEP?
M: Not me alone, but during my term the Parliament created the European Convention and achieved the minimum consensus for a draft treaty for the establishment of a Constitution for Europe. It also expanded the EU by the accession of ten new member states from Central and Eastern Europe and backed a new international peacekeeping strategy. I was also active in the networks of the Commission, the Council, and the Parliament on issues like the Brenner Base Tunnel, the proportional representation of South Tyrol, and the problems faced by mountain farmers. And I traveled to Kashmir, Tibet, and Iraq to promote the EU position on peacekeeping issues.

But it's hard to gain support for visions of sustainability in the EU, unfortunately. That doesn't mean that I am disenchanted with it all, however; if anything, it has made me more determined to come to grips with sustainable projects on a smaller scale and on my own initiative.

H: Was there anything about the European Parliament that ran contrary to your ideals?
M: The pompous blowhards who pretend to represent the interests of their voters, the lobbyists, or even Europe but are really just opportunists, trading political offices for favors, and the mad egotistical scramble for list places for the next elections—I loathed all that. As if they didn't have anything better to do.

H: You have described yourself as a political pragmatist. Did you tangle with your own party over any concrete issues?
M: I didn't belong to any party. In 1999 the Italian Greens invited me to stand as a candidate on their party list for Northern Italy. I agreed with the understanding that, were I to be elected, I would remain nonaffiliated. As an MEP, I was part of the Greens/European Free Alliance group. (My enthusiasm for the group grew when Daniel Cohn-Bendit took over the leadership in 2002.) So there could be no question of tangling with my

party over specific issues, and as a newcomer to the Green parliamentary group, I tended to hold back a little.

H: How do you see the future of South Tyrol within Europe?

M: With the introduction of the euro and the Schengen Agreement, South Tyrol is less a part of Italy and more a part of Europe. This means that South Tyrol can now develop its own European identity. I personally think of myself as a South Tyrolean, a European, and a citizen of the world, not as an Italian, Austrian, or German. European integration has enabled us, fortunately, to shake off our national identity—this was one of the reasons for my foray into European politics. However, things aren't as easy for every region of Europe as they are for South Tyrol, and Europe will continue to be a house of many nations until we finally understand that we only have a chance if we all work together.

European integration is an opportunity for South Tyrol. What we now have to do is address some important domestic issues. The ethnic-minority situation must be redefined, under the auspices of the EU, and our home region needs to be repositioned in a global context. Our status as an autonomous province brings many advantages, but no single ethnic group has exclusive rights to the benefits. Alexander Langer spoke of "the importance of the mediators, bridge builders, wall jumpers, and pan-European commuters" and of our need for "traitors of ethnic unity," but not "defectors."

H: Do you believe that you have upheld Langer's legacy in a way he would have approved of?

M: Alexander Langer was a visionary and an idealist, who unfortunately always provoked a negative reaction from the majority party in South Tyrol. In the end it was his high expectations, of himself, and humanity as a whole, that broke him. He did not fail. As a pragmatist, it was never my intention to step into his shoes and continue his legacy. However, some of his visions have now become reality, and his "Alternative South Tyrol"—more tolerant, fairer, greener—is stronger than ever.

H: Have you spoken recently to his mother or to any other members of his family about this mission?

M: Langer's mother is dead, and after his death I lost contact with his brothers. I do come across young people who are eager to take on his causes more and more frequently, though.

H: Why don't you run for election again?
M: The opportunity I had as a Green MEP to promote sustainable policies across Europe, and particularly in the new member states, was a big challenge, for sure. But I have no desire to become a career politician and spend my time glued to a chair. My museum project now takes up most of my time and energy—and my resources—and I am also eager to promote a few aid projects in the Himalayas. A second term as an MEP wouldn't allow me to do all that, because if I were elected, I would have to take on more responsibilities.

Having said that, there are some things I miss about being a politician: the opportunity to campaign for a free Tibet, autonomy for Kashmir, a different approach to agriculture in mountain regions, and sustainable tourism in the Carpathians, for example. As a normal citizen, what I have to say has less impact.

H: Can you see yourself being politically active in another sphere at some time in the future?
M: Not in the immediate future, no. If I were ever to go back into politics, it would be at a local level, in South Tyrol. "South Tyrol in Europe" is a challenge that would be worth coming to grips with.

H: During your five years as a politician, do you think that you managed to diminish the significance of the Ebners in the South Tyrol to some extent and promote a more modern version?
M: Fortunately, there has never been an "Ebner South Tyrol." However, the "Christian Brethren," as I call them, still play a key role in the relationship between politics and the media, without ever demonstrating any leadership qualities. And the virtual monopoly enjoyed by their newspaper allows them to play the individual political factions against each other.

For many South Tyroleans, civil courage, vision, and directness are alien concepts. Marginalization is the main agenda, and subservience is a duty. The South Tyrol of the "Christian Brethren" can only be prevented by separating media power from political power. As long as the gentlemen of the largest newspaper in the region get to determine who becomes president, TV editor, or mayor, as long as politicians allow themselves to be blackmailed and provincial parliamentarians subordinate their vision and ideals to a power-hungry family of entrepreneurs, South Tyrol will remain a long way from Europe.

I am powerless in this respect—labeled as persona non grata by successive generations of the Ebner family—but not without hope.

H: Are there any parallels between climbing and politics?
M: No. Climbers experience the consequences of their actions immediately and on a personal level. In a representative democracy, politicians make decisions on behalf of everyone and are generally no longer in power when the consequences of their decisions become apparent.

H: Which European politicians did you get along well with during your term as an MEP?
M: Daniel Cohn-Bendit, a politician with passion; Friedrich Gräfe zu Baringdorf, with his practical approach to agricultural policies; Elmar Brok from the EPP [European People's Party] faction, whose overall view of things is something I admire; Johannes Voggenhuber, Austrian and passionate European; Jo Leinen . . . But I also have total respect for Fischer and Prodi.

H: What have you learned from your stint in politics?
M: How difficult it is in a chaotic, globalized world to achieve workable compromises, and how important every individual vote is—in elections and, above all, in discussions about our future. I will continue to play my part, even if it is only as a lone voice in the wilderness.

CHAPTER VI
THE LEGACY

1998–2008

An exhibition must be more than just an information event. It should move the viewer, like a piece of expressionist theater.

—Chris Dercon,
Museum curator

THE POWER OF ART

The Messner Mountain Museum is located in the middle of the Alps, in South Tyrol, on the threshold between north and south. It is a place to pause and consider, an exhibition space for the mountains and their history, a place for dialogue, for the common legacy and collective biography of mountaineering.

Could there be any better place for the centerpiece of the museum—a place that addresses the subject of man's encounter with the mountains—than Sigmundskron Castle, perched on a hill with the mountains of South Tyrol as a backdrop? To the east there is a view of the Schlern; to the north, above Meran, is the Texel Group; and directly below the castle lies Bozen, the provincial capital of South Tyrol, the "Land of Mountains."

Around the central museum, MMM Firmian, at Sigmundskron, are grouped four satellite museums devoted to individual themes: MMM Juval, a castle in the Vinschgau dedicated to the "Myth of the Mountain" and the religious dimension of the mountains; MMM Dolomites, near Cortina, the "Museum in the Clouds," where the focus is on the vertical world of rock climbing; MMM Ortles in Sulden, where the themes are snow, ice, and glaciers; and the most recent addition, MMM Ripa in Bruneck Castle, which tells the story of the mountain people and their heritage.

In my museums, the relics, the testimonies of the mountaineers and philosophers, and the works of art should communicate with each other and provide information without the need for further explanation. In this way the visitors, the exhibits, and the place of encounter engage in an open, interactive dialogue. As the initiator and curator—although I wish to remain in the background—I want to present a dynamic image of the mountains, not some frozen reality, and to foster a creative space between mountain and observer in which changes of perception can occur.

I define myself as a seminomad, so my world consists of transient places, where being at home is not possible. I need a place of refuge in South Tyrol, a safe haven to return to from the high mountains and the endless ice, where

I can rest and prepare for my next encounter with cold, exhaustion, and desperation. In the mountains, the routes I take do not follow visible lines; they are just lines in my head. Wilderness, rock, and desert are transitory places that allow art to be created that equates to our projection, but they can never become our home.

On my map, the location markers are all about exposure, remoteness, space-time, and self-reliance. Visual mementos also play a part. Each trip is like a lifetime in itself, like being out there on another planet. The higher I climb, the deeper I understand my fears; the bigger the mountain, the clearer the view I have of my own existence. It is this message that my museums seek to convey.

By going to places where I do not belong, I experience the art of living— orientation through disorientation. All the deserts of the world lie within us, after all.

H: You describe yourself as a seminomad. What is that?

M: A true seminomad is someone who travels around from a winter base. In the spring, summer, and autumn months, he roams from place to place with his family, his animals, and all his worldly possessions. He goes wherever he can find pasture for his livestock. These seminomads are found all over the world: in Tibet, in the Sahara, in East Africa.

In a broader sense, this seminomadic existence has also found its way into our civilization. We have a safe haven somewhere from which we travel the world. Well, some of us at least. We are modern-day seminomads who travel by airplane.

H: At the start of your career, you were traveling all the time. When did you realize you needed a place of your own to call home?

M: I've always felt a strong desire to have a permanent base, and I was quite young when I built my own house in Villnöss.

H: Your first house was a log cabin on the Gschmagenhart Alm.

M: A beautiful place. The most beautiful place in the Dolomites. It had been a long-held dream of mine to have a hut high up in the mountains between the forest and the cliffs.

H: What is it about the Gschmagenhart Alm that you find so fascinating?

M: The Alm is a mountain pasture, a meadow, maybe eight hectares in size, with the forest below and the impressive sculpted peaks of the Geisler as a backdrop. I spent my summers up there as a child, with my brothers. When we didn't have to look after the chickens, that is! We slept in the straw, fetched water and wood for the fire, and went climbing. There was no electricity, no telephone, no shop, only what we carried up there.

For about ten years I dreamed of having a place to live up there. I finally managed to buy two old huts, which I tore down, and together with a couple of carpenters I built a little cabin. It had a kitchen with a stove, a living room with an open fire, a tiny toilet, and a bedroom. It was the simplest house you could imagine, but it had a view of the Geisler.

H: You didn't live in your dream home for long, though. Why was that?

M: My idea was to have a permanent place to live up there, somewhere I could stay when I wasn't traveling around like a nomad. But I had to earn money, and that meant I needed a telephone and a little more mobility. Back then, in 1971, there were no cell phones and in winter it was freezing cold up there. Living on the Alm just wasn't practical.

It's still a lovely place, though, precisely because it is so simple. If it were a normal house, it wouldn't have the same feel to it.

H: You then bought a former rectory at auction, didn't you?

M: Yes, I did. Years ago, the farmers in the valley built the parish priest a house to encourage him to stay. But he left anyway and went to another valley. The house later became an old folks' home and then a school. At the end of the 1960s, the place was shut down by the provincial health authority, because the toilet facilities and windows were not up to standards and there was no running water. In 1972 it came up for auction, and I bought it.

H: For 6000 deutsche marks?

M: For 6 million lire [US$10,500], yes. In 1973 the two of us moved in—my wife, Uschi, and me.

H: And how was the seminomadic life at that stage? Was the house just a place you could escape to, to rest and prepare yourself for your next adventure, or did it feel like home?

M: It was my home. It was like a nest. Incidentally, my parents also lived in the valley, so I could visit my mother, and go to the village shop for groceries and the newspaper. I couldn't imagine ever moving away again. The only limiting factor about Villnöss was the fact that I wasn't able to buy a farm, which was a dream of mine. Even in those days, I dreamed of being self-sufficient. I wanted to be a farmer.

H: You dreamed of being self-sufficient?

M: Yes. Being seminomadic means being independent. It also means doing everything yourself. But without a farm, I couldn't do that.

H: Where is the satisfaction in that? Surely you could just go and buy your ham from the shop?

M: I did eventually realize that it's actually cheaper to buy wine, ham, and bread from a shop. And it's also cheaper to stay in a hotel than pay for the upkeep of your own mountain hut that you only use for two weeks in the summer. But having a place that feels like home, somewhere I could imagine myself living for the rest of my life, somewhere to come back to, was much more than that. Maybe it was a desire to put down roots. And I wanted to grow my own food, to be independent of the outside world—of success, income, a pension—until the day I died.

My dreams of self-sufficiency might also have something to do with the fact that the Villnöss Valley is a farming community. The farmers decided everything; they were the kings of the valley—and rightly so. They were the custodians of the cultural landscape and the local way of life. You could become firmly embedded in that society, and have a measure of security, only if you owned land.

H: There is a paradox here. When you are away on expeditions, you willingly put yourself in life-threatening situations, yet at home you want maximum security. You want to keep your own pigs and produce your own wine and milk. You want to have total control over what you eat and the basic essentials of your life.

M: Self-sufficiency gives you the greatest possible security. That applies to all of us, particularly today. This kind of security can't be taken away from you, unlike paid work or a state pension, which can shrink over time. The flip side is that you have to take greater responsibility for yourself and you need the know-how, the skills, and the energy. You only get out of it what you are prepared to put in. You can only reap what you sow.

I believe that, in Europe at least, we have made the mistake—politically speaking—of taking too much responsibility away from people. Citizens who once took responsibility for themselves have become dependents, demanding more of everything—subsidies, incomes, pensions. When the responsibility is taken away from them, people's aspirations dwindle, until finally we reach a state of zero growth and begin to descend into social chaos, a state of limbo. Welfare democracy is a dead-end street. What I always wanted was maximum responsibility for my own life. I've always had high expectations of myself. Others call it ambition.

H: What does the concept of home mean for you?
M: Home is where my children are and a place where I can put down roots.

H: So is Villnöss home?
M: It was, yes. That's where I grew up. That's where my family was, a family that still stands together to this day.

H: Are the Himalayas home?
M: No, no. Well, maybe K2 Base Camp for six weeks.

H: But you have brought back a great many things from the Himalayas that you now have in your houses. Why is South Tyrol your home and not Nepal?
M: Nowadays the place that feels most like home is Juval, a place I have created something. For me, the idea of home has very little to do with my childhood memories, the grave of my parents, or the mountains of my youth. Home is no longer the cozy homeyness of that background world.

H: Could you ever feel at home in a rented apartment?
M: I couldn't, but others can and do. Home is a relative concept.

H: So an entry in the land registry is a prerequisite for you to feel at home?

M: It helps, yes. I have to own a property if I want to make it the way I want it. So having my name on the title deeds is important, I suppose.

H: Coming back to the rectory, when you moved in with Uschi Demeter was she happy with the house, or did you make the decision to buy it more or less on your own?

M: No, she decided, not me. And she still lives there. It's her place, her responsibility; she owns it. She supervised the renovation and furnished the house and designed the garden, after we'd talked it all through together. I was in the Himalayas when the main work was being done.

H: How did the local community treat the new arrivals at the rectory? After all, you were someone who had made climbing his profession, you lived with a woman you had not married in a church, and you were also on record as saying that the South Tyroleans had a problematic, even hypocritical, relationship with their past.

M: People's reactions were different. Those who knew me liked me, but most of them thought I was a bit strange. In fact, I was just like them. Their respect for me started to grow only when I achieved financial success. They didn't recognize my mountaineering achievements, but when they saw that I was getting by financially, that was okay with them.

H: How long did you both manage to put up with the situation at the rectory? There's Uschi, waiting at home for you, you get back from your trip, she's pleased to finally have her man home again . . .

M: And the next day he tells her about his new idea for an expedition. The day after that, he goes training, and then he starts buying equipment, making plans, and working out how to finance the trip. And three months later, he's gone again.

H: So you never just sat there in the autumn sunshine, looking at the cows and telling yourself how nice it was to be home again?

M: That was ultimately the reason why we split up. She wanted to go to places like New York, Paris, and Provence, visit art exhibitions, that kind of thing. But I needed all my time and money for the next expedition. It was unfair and selfish of me, for sure, but that's the way my life was.

H: Why didn't you and Uschi start a family? Were you scared of doing that or just restless?

M: I couldn't imagine myself looking for a job and having a summer vacation at the seaside and a week skiing in winter. I still had so many projects in my head—Mount Everest and a lot more besides. Having a family would have been too much responsibility. The desire to have children came later. I come from a big family myself, after all. I think people who come from big families tend to want to have children of their own.

H: Did you get harassed by tourists when you were living in Villnöss?

M: I lived up on the hill above Saint Magdalena, I was a well-known man, and I wrote books, so people were curious. The locals soon realized that the tourists wanted to see where Messner lived. It was good for business.

H: Did you get busloads of them arriving at the house then?

M: Sometimes. But I couldn't spend all my time meeting and greeting guests. It was my home, not a hotel.

H: So they wanted to look around the house?

M: Some of them, yes. Finally, I had to lock the door. That was one of the reasons I didn't make many friends in the valley. When you live in a mountain village, you can't just lock your door and shut people out. People expect to be able to turn up unannounced. That's always been the accepted custom there.

H: So people used to just open the door and wander in?

M: Thirty years ago in Villnöss, you'd go out for a walk and, if you felt like it, you'd just wander up to one of the farmhouses, knock on the door and say hello, and ask them how they were. And if they had the time and they were in a good mood, as often as not they'd invite you in for some bread and ham and a glass of wine. When I went for a walk with my mother, people would often ask her in and they'd end up chatting. That's the way people communicated in the mountain valleys. Hardly anyone read a newspaper. They met and shared information, thoughts, memories, and emotions.

I think that's a nice way of doing things. But when you get hordes of tourists on your doorstep, that's a different matter. I had neither the time nor the inclination to pour every vacationer a glass of wine and tell them how I climbed the north face of the Eiger, or Nanga Parbat. They all wanted to hear the same thing, and they all asked the same questions. They could have just gone to one of my lectures. Back then, I was doing lectures for two thousand people in places like Munich, Milan, and Naples. I got to most big cities at one time or another. But the summer guests didn't want to do that. They wanted . . .

H: To sit and drink wine and eat bread and ham with you in your front room.
M: Yes. At the time I was also doing a lot of talks for climbing clubs and organizations, like the German and Austrian Alpine Clubs. Part of the arrangement was that after the event you sat with the committee members and told the whole story again. It sometimes went on until three in the morning. I had to give up all that cliquey, clubby stuff; otherwise, I'd have turned into an old windbag myself.

I can't stand all that pseudoprivate chummy nonsense. And I don't do it anymore these days. Public and private are separate things now. I do a public appearance, and afterward I'm gone. These days, after a public lecture I go straight back to my hotel. I wouldn't think of staying at someone's house to relax and wind down.

H: So you became too big for the narrow confines of the Villnöss Valley?
M: That's a total misconception. But it's exactly how people interpreted my behavior. I was a walking advertisement for the valley, a tourist magnet. I told people that even if I didn't let people get up close and personal, I was still an asset for the region. But it was my home as well, and I just wanted to be left in peace. I tried to get them to understand that I was only ever there for a few weeks at a time, writing a book, working and preparing for the next expedition.

H: Did the locals in Villnöss like the Tibetan prayer flags that you had flying from your roof?
M: They did, yes. And the statues in front of the house. My passion for collecting things was so great that all the cellars were soon full of stuff—

Tibetan artifacts, rustic furniture, and pictures. Even then it was clear that sooner or later I'd have to find somewhere to put it all. In 1979 I started looking for a castle.

H: Why did it have to be a castle? Did you want to barricade yourself inside it?
M: I was looking for a kind of eagle's nest, the perfect place of residence for a seminomad like me. Somewhere people couldn't just knock on the door and come in without letting me know first, somewhere no one could see me wandering around inside. I wanted some peace and quiet.

At Juval, the medieval castle I finally bought, no one can catch us unawares—no fans, no guests, no political celebrities. And it's impossible to see into the bedroom from below, from the valley, even with binoculars. Juval Castle is perched high up on a cliff, a long way from anywhere.

H: Your father was against you buying Juval, wasn't he?
M: Yes, he never visited me there.

H: Did he say why?
M: He said that Juval wasn't a suitable place for me to live. He was right, too. A South Tyrol climber without a proper job shouldn't live in a castle.

H: Did your father think you living there was immoral?
M: He saw it as arrogance on my part. He was also worried that I might have bitten off more than I could chew, that I might squander all my money. He had helped me find the house in Villnöss and supported my decision to buy it, but in his eyes buying a property like Juval was beyond my means. He disapproved of that kind of thing. He thought I had ideas above my station. And as I said, he was scared that Juval would ruin me.

He could have looked at it another way, I suppose—the lad will have to stay at home now and stop taking off on those expeditions once and for all. But I couldn't subscribe to those bourgeois attitudes. Buying and renovating Juval was a challenge, a game. I still live there for part of the year.

H: And your mother?
M: My mother liked Juval.

H: Were you together with anyone when you bought Juval?

M: I was single at the time. A friend showed me the place. It was practically a ruin.

H: How much did you pay for it?

M: Sixty thousand marks [US$35,000]. That was a lot of money for me then, but it was financially feasible. I worked on the place for two years, then the previous owner sold me another two farms. He had money, but he was so cheap that he'd let them fall into disrepair. Which was lucky for me in a way, as I wouldn't have been able to afford to buy them otherwise.

H: So he was too stingy to renovate them?

M: Yes, if the roof was broken, then it was broken, simple as that.

H: What state was Juval in when you bought it?

M: Pretty dilapidated. There were piles of rubble 2 meters high, some of the walls had collapsed, and bits of the roof were missing.

H: Why did you decide to take it on in the first place? The first time you went there, you had to climb a 6-meter-high wall to get in, didn't you?

M: Yes, and inside the walls I saw Himalayan cedar and boxwood, and there were [Tilman] Riemenschneider frescoes in the dining room. I had no idea how they got there, but I liked the place right away. I fell in love with the castle, the location, the ambience. It wasn't a gloomy castle; it was light, elegant, and full of warmth. I figured that with a bit of work, it would be the ideal place for me.

The first time I went there was on an autumn afternoon, and the sun was shining. I immediately thought, "That's my castle!" It was like it had been built especially for me, eight hundred years ago. I acquired the farms around it a couple of years later, and the dream was then perfect. I had my farm, my self-sufficiency, my castle, and soon a family, too. I imagined myself living like the Russian landed aristocracy of two hundred years ago—summers in a castle, surrounded by my land and my animals, and winters in the town.

H: When did you purchase Juval?

M: 1983. At first, there was only one room that was habitable—where the living room is now. That's where I camped, with a sleeping bag and a mat, whenever I was in South Tyrol. I helped with the renovation work.

H: You once said that Juval suited your mood and your imagination.
M: Juval is better than I ever imagined my dream home—my "Castle in the Air"—to be. The location alone is stunning—a Renaissance palace in the mountains, the most beautiful castle in the Alps.

H: So why did you need the farms as well, your "Noah's Ark," as you call it? Wasn't the castle enough for you?
M: The farms were actually more expensive than the castle itself. And it's hard to describe just how run-down they were. The Oberortlhof farm didn't have a single intact roof. You could see the sky from the cellar. And everything was overgrown, with brambles everywhere. I had to get someone in to cut it all back; otherwise, the grazing land would have been ruined. There was nothing there; I had to buy everything new, machinery and so on.

H: Did you get a sense of quiet satisfaction from owning this castle and therefore showing the South Tyroleans that you didn't just climb big mountains but were moving up the social ladder as well?
M: Things like that don't concern me. And no one else apart from me would have wanted the place anyway. To pursue a career in the accepted sense, all I would have had to do was conform to what society expected. But I didn't want to do that. In fact, back then I hadn't discounted the possibility of moving away from South Tyrol completely. It was the idea of a castle in the air that was important, not representing myself to the outside world.

Besides, the farm was just one big problem at first. I turned to Luis Durnwalder, the provincial president, for advice about how best to organize the farming side of things, and he told me exactly what he thought about it: "Buying a farm like that is just stupid. You are a climber; you're never there. And if you have to employ people to run the place, it will just be a lot of trouble and debt." How right he was! He gave me a piece of advice: "There's also an area of woodland for sale that used to belong to the castle. It will cost you about the same as the farm." I was astonished that he knew so much. "It's up to you, of course, but the woodland will only increase in value, and there are hunting rights that go with it." Durnwalder was absolutely right.

I went home, thought about it, and decided to stick with the farms. If Durnwalder was annoyed, he didn't show it. "The only thing you can do now is make the best of it," he said, "I'll send you my best people. They can advise you." He came to see me and brought along an apple expert, a wine expert, and a livestock expert. All they asked for in payment was a bite to eat.

H: Did he laugh at you when you told him about your plans for an organic farm?
M: He said it was a unique idea but it wouldn't work, that organic farming was not possible, not anymore. I did it anyway, apart from the wine. I now have an organic farm, a winery, and some fruit trees. These days, if he happens to come visiting, he always asks me if I'm still managing to survive financially. What he said was right, though; I'll give him that.

H: What do you tell him?
M: That we're surviving. We've got a few problems, but we're surviving.

H: You once said that a sedentary life would sooner or later lead to lethargy and depression.
M: Yes.

H: Has that been the case at Juval?
M: I've never reached the point where I might become lethargic. Juval is still a work in progress. There's always something that needs improving. Then there was the idea of turning Juval into a museum, a museum that pays for itself.

H: Was the redesign and development of Juval more important to you than living there?
M: Juval has become a kind of theme park, a *Gesamtkunstwerk*, a total work of art, and part of my museum concept.

H: Do you enjoy sitting in your castle garden in the summer, or do you get restless and start thinking about things you could still do to improve the place or your next trip away?
M: I really enjoy evenings at the castle, especially when the kids are there

and we eat outside or have a barbecue. It's nice to spend time together. But I don't sit there and think, "That's it—done." It's good how it is, but it's never enough. My imagination gets fired by new ideas and new possibilities, not by the satisfaction of having meat from my own farm on the barbecue.

I enjoy drinking a glass of good wine from my own winery and eating the produce from the organic farm, of course—the jams, vegetables, and home-produced sausages. And most of all, I enjoy being with my family. I like the view of the woods and fields, the farms, and the pigs running around, but that's not enough on its own to sustain me emotionally.

Having everything is boring; I'm convinced of that. Once you have something—knowledge, skills, possessions—or have achieved something—climbing Mount Everest, for example—it becomes banal. The mountain is only a challenge before I climb it. That counts more than the success afterward. The experience remains, for a while at least, but for me the feeling of curiosity about the next challenge, the next question, is always stronger.

H: So even Everest was boring after you had climbed it?
M: Yes, after a few weeks that's the way I felt about it. By then, I was already on another trip—to solo Nanga Parbat.

H: So you are interested in these things only for the time it takes to climb them, is that it?
M: When I am identifying an objective, the summit is everything. I wouldn't make it up there otherwise. I'm not a superman; I'm just mentally capable of concentrating on the end point.

Once this has been achieved, I need a new task, a new idea, a new project. I've been lucky so far—I've always been able to get myself motivated for the next new thing. The challenges I set myself are age-related. I don't rule out the possibility that with my museum project I might have found a challenge that will take me into old age and keep me from getting depressed.

H: That makes it sound like you are manic-depressive.
M: No, I'm not like that. On the contrary, doing things gives me a real zest for life.

H: You get all fired up for an idea, and then, after you have achieved your objective, you sink into boredom and depression.

M: No, it's not depression. It's more like taking a deep breath and realizing how valuable life is. But only if I live it to the fullest, only if I participate.

H: You are often accused of being egoistic, and you also describe yourself as an egoist. Is building several museums in South Tyrol part of that egoism?

M: I didn't build them for South Tyrol; I did it for me. I set up the museums because I enjoyed doing it, and because I felt I had to. I had to climb Mount Everest, and I had to set up the Mountain Museum project. It's my way of expressing myself.

I've written books, climbed mountains, and trekked across deserts. I now feel the time is right to put everything I know—and everything I don't know—into the museum and express myself in that way. If we are enthusiastic about what we do, we don't harm anyone. It's the people who always need an excuse for their egoistic behavior and demand recognition for what they do that I have no time for. Idealists of that type are usually unhappy with themselves and the lives they live.

H: Why is a museum a better way of expressing the nature of the mountains than a film?

M: I am more interested in human nature than in the mountains themselves. And with the museum, I have more means at my disposal—art, texts, music, and sounds—and more possibilities than with any other medium.

I'm primarily concerned with what happens inside a person when they encounter the mountains. When you climb a mountain, you come back down as a different person. We don't change the mountain by climbing it; we ourselves change. Not much happens to the mountain itself—unless we build huts, trails, roads, ski lifts, or other infrastructure. We might leave a few crampon scratches, but that's about it.

But plenty happens to the person. Up there is where all the masks fall. And even after the climb, a fellow climber will often show his true face. Envy, jealousy, and greed frequently present themselves only when the self-proclaimed good companion is back among his peers. This truth, together with the history and the shared passion, is what my museums seek to reflect. The idea is that, ultimately, everyone benefits from it—the climbers and their critics, the art lovers, everyone who visits. As a region for tourism, South Tyrol benefits, too—and so do I, of course.

The parallels are obvious. I have always said that climbing is an end in itself. There is no other reason to climb mountains, no other motivation, than your own passion, ambition, and love of nature. My personal aspiration to do it as well as I possibly could, or even to be better than others, is something I have taken with me into the museum project. I have always tried for quality in everything I do. I don't have a problem with that; I have a problem with mediocrity.

It is only because I have the courage to stand by my ideas, my projects, and my aspirations that I am frequently branded as an egoist. I've yet to meet a person who isn't an egoist. Picasso was a magnificent egoist, and he painted his pictures because that's what he had to do. But I believe that, as egoistic as he was, Picasso created something for the human race that can never be repeated. The world would be a happier place if only there were more people capable of expressing their vision, ideas, and needs.

H: How did you first come up with the idea for the museums?
M: Like most things in my life, the idea developed from something I was passionate about. When I wrote my essay about free climbing called "The Murder of the Impossible," one of the sources I found most inspiring was from a climber by the name of Paul Preuss.

After the piece was published, an old lady with scrawly handwriting I could hardly read wrote me a letter. It turned out to be a real eye-opener. In the letter she waxed lyrical about her young love, Paul Preuss—a Viennese Jew, educated, affable, incredibly shrewd. Preuss fell to his death in 1913. The old lady sent me another few letters—chatty stuff mainly—and a hammer, Paul Preuss's climbing hammer, together with the request that after my death the hammer be passed on to a like-minded person or be made available to the public.

The hammer lay in my cellar for a long time, but I knew it couldn't stay there forever. Then, in the 1990s, when I was recovering from my broken foot, I started developing the idea of the tiny little museum in Sulden, "Alpine Curiosa." I bought a little hut, the "Flea Hut," in Sulden, developed a concept for the place, and, together with Christoph Ransmayr, drafted some texts to accompany the exhibits. The museum tells its story using a rock-climbing feature—the crack—as a metaphor for the gap between what the great mountaineers wished to achieve and what they actually did.

Preuss, for example, wanted to climb without the use of pitons, but the fact is, Preuss did place pegs, even if was only twice in his entire life. That little weakness is what makes him human. Preuss is my idol, but he always had his little hammer with him when he climbed.

My Alpine Curiosa mini-museum is open summer and winter, and entrance is free. I've basically given it to the village community. It was a lot of fun collecting the exhibits, designing the museum, and writing the texts—a bit like doing a book, in a way.

One of the items on display is a piece of [Edward] Whymper's rope, the one that broke on the first ascent of the Matterhorn, killing four of his companions and leading to the accusation that Whymper had cut it to save himself. Next to it is the ice ax that belonged to Toni Egger, who fell to his death on Cerro Torre after his and [Cesare] Maestri's purported, and since discredited, first ascent—another good story.

There is also a piece of the cross that was placed on the summit of Mont Aiguille in France after the first ascent five hundred years ago. And, next to that, a bolt and chisel belonging to Emilio Comici, who in 1933, before he climbed his magnificent new route on the north face of the Grosse Zinne, wrote that it was acceptable to take normal rock pitons, carabiners, ropes, slings, and etriers with you on a climb but not to drill holes in the rock and place bolts. Yet he still took the chisel with him. It was found by the climbers who made the second ascent. I had to beg for five years to get that chisel, a piece of alpine history.

What else have I got? The book about the heroes of the first ascent of the north face of the Eiger in 1938, one chapter of which ends with the words that Heinrich Harrer is reported to have said to Hitler after the climb: "We have climbed the north face of the Eiger, beyond the summit, to you, our Führer." I'm not sure how that actually works in practice—how do you climb beyond the summit?—but that sentence certainly helps us to understand our history. Also—and this is the point—Harrer always disputed his association with the Nazis and Hitler.

H: The main museum is near Bozen at Sigmundskron Castle, a national landmark and an important political symbol in South Tyrol. Why there of all places?

M: Over a period of some four hundred years, the castle had fallen into decay, and after a legal battle, the ruins were finally acquired by the provincial government. There were piles of rubbish outside, and no one really knew what to do with the place. I put forward my suggestion and was granted the use of the castle for a fixed period of time. The castle is right next to the autobahn. It is totally unsuitable as a residence; it could only be used as a museum.

I began to invest some money in Sigmundskron, bought out the local innkeeper who ran a restaurant there, commissioned various artists, and drew up some plans. All of a sudden a campaign was started against me, initiated by the two brothers who control the media in South Tyrol, Toni and Michael Ebner. It was an unprecedented campaign, full of malice and speculation. To this day, I still don't understand why they did it.

H: Hadn't you done some books for the Ebners' publishing company in the past?

M: That's right. Their father published a few of my early books. I think I was one of their most successful authors back then. My relationship with the Ebners' publishing house was businesslike rather than friendly, a partnership of convenience.

It was after the Everest expedition, when I was asked why I hadn't carried a flag to the summit and I told them my handkerchief was my flag, that the Ebner media began to take potshots at me. The Ebners' daily newspaper, *Die Dolomiten*, violated all the boundaries of fairness. They slaughtered me in the media for weeks on end, trying to discredit me. It got so bad that I went to see them and told them I wanted all the rights to my books back, whereupon Toni Ebner Sr. told me that I'd never make a living in South Tyrol without his backing. I left after that and went and found another publisher, outside South Tyrol. I received my formal discharge from Ebner in writing a short while later.

H: So you and the Ebners had an old score to settle. Did you get the feeling that they wanted you out of South Tyrol?

M: I kind of got that impression, yes. In any case, they knew I'd be staying if I got the go-ahead for Sigmundskron.

MMM Firmian at Sigmundskron Castle near Bozen in South Tyrol. Opened in 2006, it is the centerpiece of the Messner Mountain Museum.

H: How did the Ebners try to disrupt your plans?

M: Even the provincial president was put under pressure. It was in the local papers. It was reported that the Ebner brothers had called the president and asked him to come to their office. He duly arrived at the publishing company, where he was told that I must not be allowed to proceed with the Sigmundskron project and received the following warning: "We're going to shoot down the Sigmundskron project, and if you don't get out of the line of fire, we'll shoot you down with it."

That statement, which appeared in huge letters on the title page of the South Tyrol weekly magazine *FF* was later denied by the Ebners. The provincial president also denied that those were the exact words used.

H: Apparently, there was also an opinion poll in which 70 percent of South Tyroleans said that Messner should not get Sigmundskron.

M: That's right. And the results of the poll are actually accurate! However, you need to bear in mind that the question was asked in a very suggestive manner and I had also been hounded by the media for months. What kind of question is this anyway: "Do you think it is right that Reinhold Messner should be given money from the provincial government and receive Sigmundskron as a gift?"

I didn't receive Sigmundskron as a gift at all. The fact is, I took on the obligation to run the place for thirty years, using my own resources and without any state subsidies.

H: What happens after the thirty years are up?

M: Sigmundskron reverts to the province, together with everything I've invested in it, although I'll be allowed to remove my exhibits, naturally. But my ambition is to make it so good that South Tyrol will be only too happy for my grandchildren to stay on and run the place.

I've told my children that in thirty years' time, if I'm no longer alive and they have successful careers of their own and don't want to keep looking after my museums, they should give the whole thing to the province. It's not worth selling.

H: Nevertheless, 70 percent of German-speaking South Tyroleans said it would be better without you.

M: It may well be more than 70 percent who are against me—no wonder, when public opinion is inflamed like that week after week. But I'll survive.

H: And in spite of all this, you still want to do something for the region?

M: Yes.

H: Why?

M: I am a South Tyrolean.

H: They are against you, and you want to give them museums?

M: I grew up in this land, so it's my land, too. I do all this for my children, too, so they can create their home here.

There is a growing feeling that there is a need to develop ideas in areas where the all-determining People's Party has no idea which way to go. I know that this region has to position itself in a globalized world. And I am a part of this region. We occupy a key position in Europe—a fantastic opportunity that must not be squandered. Our provincial president has done some excellent work, but he struggles sometimes, because he doesn't always defend himself against the populists, the monopolists, and the folksy homespun associations interested only in their membership numbers or their businesses.

I am a South Tyrolean, and I am staying here, even though I was often at the point of telling myself it was time to go. In the context of the museum dispute, I visited Trento and Innsbruck quite regularly. Both places offered purpose-built properties to house the museum, on hills in the Etsch Valley and the Inn Valley, in the event that I was refused permission to use Sigmundskron Castle. I could have found a site for the museum anywhere, really, but the best place for it was, and still is, South Tyrol. So that's what I fought for.

H: Why do you need so many museums? Isn't one enough for you?
M: My children keep asking me the same question. There simply isn't enough room at Sigmundskron. It's a beautiful site, with plenty of space outside for big sculptures, but there isn't room for the pictures on the themes of "Rock," "Ice," and "Mountain People," so I have satellite museums with Sigmundskron as the centerpiece. The idea is that the experience encompasses the whole of South Tyrol.

H: The Deutsches Museum in Munich manages to keep everything under one roof.
M: Of course they do, because they have the space for it. The Deutsches Museum is much bigger, and they have huge storerooms as well. Ninety percent of the treasures are stored in the cellars. And they require millions in subsidies—understandably—to survive.

I don't receive any funding; my museums have to be self-financing. The "extreme places" I've chosen as museum sites are also part of the overall concept. The MMM Dolomites museum is perched on a 2181-meter-high lump of rock in the middle of nowhere! It's worth visiting for the location alone. And it's the perfect place to showcase the theme of "Rock."

H: What is the overall concept behind the museums?

M: The museums are dedicated to the mountaineering legacy. They tell the story of everything I have experienced in the mountains—through works of art, relics, and quotations.

MMM Firmian at Sigmundskron Castle near Bozen addresses the subject of man's encounter with the mountains. Since the buildings that comprise the ancient fort cannot be altered in any way—they are protected by law—the exhibition space is too small to accommodate all the themes I want to address. MMM Firmian is an interactive museum that offers music, literature, and theater events, and houses collections on alpine history and the mountains of the world.

The museum in Juval Castle in the Vinschgau is dedicated to the religious aspect of the mountains. In MMM Ortles in Sulden am Ortler, the theme is ice. And on Monte Rite, in the heart of the Dolomites—the most beautiful mountains in the world—we are introduced to the vertical world of rock climbing. The fourth satellite museum, MMM Ripa, located at Bruneck Castle, is devoted to the subject of mountain people.

The museums together comprise the Messner Mountain Museum, although I am merely the initiator and am not present in person.

H: What is on display at the ice museum, MMM Ortles?

M: At MMM Ortles we present images of ice, from ice crystals to the 4000-meter-thick ice cap of Antarctica, from avalanches to the polar sea. The mountains of the Ortler Range stand as a symbol for all the great mountains of ice, while the mountaineer, Arctic explorer, and landscape artist Julius von Payer serves as the mediator between the element of ice and a broad public.

The museum lies inside the mountain, underground, at the foot of the Ortler. The mountain itself towers directly above the entrance. Inside the museum we see snowfields and seracs, we hear the cracking of the ice and the thunder of an avalanche, but we also experience the peace and quiet that exists away from all the ski resorts.

H: Can you explain the involvement of your friend Christoph Ransmayr?

M: Reading Christoph Ransmayr's superb novel *The Terrors of Ice and Darkness* encouraged me to examine in detail what becomes of our adventures and the places these adventures take place. Christoph helped me with the texts for the Alpine Curiosa museum, and it was he who had the idea of intermeshing real mountains with representations of those mountains in MMM Dolomites. He also advised me on the design for MMM Firmian. Through our travels together and the discussions we had, we became friends. Our shared experiences in the mountains are channeled into the museums—and work on the museums encourages us to get out into the mountains. I like traveling with Christoph.

H: What do Stephan Huber's work *Shit Happens* and Doug Aitken's video installation *thaw* mean to you personally?
M: Stephan Huber's work—three mountains made out of plaster in MMM Dolomites, the avalanche installation in MMM Ortles, and the installation about the conquest of the Alps in MMM Firmian—stand as strong statements in themselves. They are positioned in the middle of the gallery space, and because of this they enhance my collection and turn it into a *Gesamtkunstwerk*. Doug Aitken's installation tells the story of water and ice, and as such it belongs in Sulden.

I don't really want to say any more about them. Artists with something to say have a mediatory role in my museums and require no additional commentary from me. I exhibit their work in a context, together with others, creating an ambience that allows visitors to access the theme of the mountains.

H: What is the concept for Juval?
M: MMM Juval has grown slowly. The castle is a museum in spring and autumn, and my family's home in summer. The collections are dedicated to the religious aspect of the man–mountain encounter, with the main focus on Tibetan Buddhism, Tantrism, and animism.

H: A central figure in your Tibetan collection is the fabled character of Gesar of Ling. What is it about this character that you find so fascinating?
M: Gesar of Ling was a religious founder and is something of a hero in the collective memory of the Tibetans. The epic story in the oral tradition is set in the mountains of Kham and the Himalayas, where the mountains are seen as a bridge to the afterlife.

H: What do you hope to convey to visitors with this unique collection of Tibetan relics?

M: One room in Juval Castle is devoted to the epic of Gesar of Ling. It's called the room of "One Thousand Pleasures," named after Gesar's favorite wife. The many other exhibits that form the Tibetan collection were collected over thirty years and three dozen expeditions and are spread across several inner courtyards and rooms. Although occupied by China, Tibet is still a realm of snow where the mountain gods reside. The museum at Juval tells that story with the help of works of art from Tibet.

H: With all due respect to your collector's passion, surely a collection of Tibetan folk art belongs in Tibet, or if not there then somewhere nearby at least?

M: These days, Tibetan folk art is scattered all over the world. The Red Guard destroyed most of the monasteries and sold off many of the bronzes, *thangkas*, and artisan craftwork. In Tibet itself there's hardly anything more to be found. The export of antiques is forbidden, by the way. I acquired the best pieces in Zurich and London, not in Lhasa, and others in Kathmandu. A lot of them come from the craftsmen in Dharamsala in India, where His Holiness the fourteenth Dalai Lama lives in exile, together with forty thousand Tibetan refugees.

I see myself purely as a custodian of a culture that was about to be destroyed. When Tibet is finally granted its cultural autonomy, the greater part of my collection will go back to where it came from before it was taken out of the country by communist functionaries, Tibetan refugees, and monks. If no one had taken stewardship of the Tibetan culture, a culture that has been seriously marginalized, it would have been lost forever. However, there are many people who love Tibet, and the Tibetan people are strong, so we may yet manage to save Tibet's cultural legacy.

H: What was the thinking behind the Dolomites museum on Monte Rite?

M: The MMM Dolomites site enjoys the best possible location, right in the middle of the white limestone mountains of the Dolomites. And it all came about by chance. During the eternal wrangling over Sigmundskron Castle, I went looking—in desperation, really—for a site for a satellite museum to showcase the theme of "Rock." And I found the old fort on the summit of Monte Rite, between Cortina and Belluno.

Unlike in South Tyrol, the media, the mayor, and the politicians in the province of Belluno received my suggestion with great enthusiasm. In only three years a little mountain museum was built and filled with my exhibits. I've been running it since 2002 without any subsidies.

H: What can visitors expect to find there?
M: All of my museums are in powerful locations. MMM Dolomites probably has the best setting, even though it's a long way from any major roads. It isn't easy to get to, but when you do get there it's a fantastic place. It's worth going for the view alone—a spectacular 360-degree panorama of the Dolomites. On a clear day you can see a hundred peaks or more.

Inside, the museum is designed with a long main gallery and twenty or so side chambers. It houses paintings from 1800 to the present day, together with fossils, land art, and video installations, and tells the story of climbing in the Dolomites, from [Dieudonné] Dolomieu to Alexander Huber, the best rock climber of the new generation.

H: Where exactly is the museum?
M: On Monte Rite in Cibiana di Cadore near Cortina d'Ampezzo in the middle of the Dolomites. It was a miracle that I found the place at all. The so-called "Museum in the Clouds" is housed in an old fort at 2181 meters on the summit plateau of Monte Rite.

It was built by the Italians in 1914 as part of the line of defense against the Austro-Hungarian kaiser Franz Joseph. Then came the First World War. In 1917 the Austrians occupied Monte Rite. A year later they were forced to withdraw, but not before they had blown up large parts of the fort. In the Second World War the ruined fort offered shelter to partisans and thereafter sank into a long, deep slumber before I discovered it in 1998. The restoration was completed within three years, a collaborative venture with the municipality of Cibiana di Cadore and the region of Veneto, and the "Museum in the Clouds" opened its doors in 2002.

Inside, the renovated fort resembles the nave of a church with twenty side altars, where relics and memorabilia tell the phase-by-phase story of the development of the Dolomites, with reference to those climbers who wrote alpine history with their first ascents and new routes. From the British pioneers to Michl Innerkofler and Georg Winkler; from Angelo Dibona, Michele

Bettega, and Gabriel Haupt to Emilio Comici, the father of "sixth-grade" climbing; from the hero of the *direttissima*, Lothar Brandler, to [Heinz] Mariacher's landmark route "Modern Times" and Alexander Huber.

But the heart of the museum is the art collection. The paintings include views of the Dolomites from the romantic period to the present, from Thomas Ender to E. T. Compton, Hamish Fulton, and Stephan Huber, whose white plaster mountain sculptures stand like erratic boulders in the Great Gallery of the museum. Between the paintings, the windows frame spectacular views of the surrounding peaks. I call them "real paintings."

And there are other relics as well that are older than the rest: the beautiful fossils from the time when the Dolomites were an enormous coral reef in a tropical ocean, and the skull of the legendary cave bear of Conturines. You can walk out onto the roof of the museum and see glass sculptures shaped like Dolomite crystals, standing on the former gun emplacements.

H: How did you find the place?

M: I found it on my birthday, September 17, just before I took up my seat in the European Parliament. I was actually looking for something completely different, but I immediately knew that it would be a great place for a Dolomites museum, although the idea was still only fragmentary at that point.

I stood up there with my wife and a few men from the village of Cibiana di Cadore, looked around at all the peaks I'd climbed decades before, and came out with the suggestion there and then—to take a hundred-year-old ruin and turn it into something for the future, then fill the place with things that relate to the Dolomites. The idea has now become reality.

H: What was the idea behind MMM Ripa?

M: The name "Ripa" is derived from the Tibetan words for mountain and man. The aim here is to show the special relationship that mountain people have with their environment. Although their religions and their art may be different, the people I've met in the various mountain regions of the world have often developed similar strategies for survival. MMM Ripa is housed in a 750-year-old castle in Bruneck, so it has its own history as well as being the perfect place to showcase the history of mountain people. Our aim was to combine the two.

H: What are your concrete plans for Sigmundskron?

M: As the centerpiece of the Mountain Museum, Sigmundskron houses paintings and sculptures from my collection and is dedicated to the history of mountaineering, from Moses and Mount Sinai to Tomaz Humar and his solo ascent of the south face of Dhaulagiri. It is these stories that define us as mountaineers; they represent our heritage. They are what the climbs of tomorrow are built upon.

MMM Firmian at Sigmundskron is the creative motor and the beating heart of the museum project. It is a venue for conferences, festivals, lectures, and concerts on the theme of the mountain. The administration for the entire museum project is also located there.

H: You describe the museum project as your "fifteenth eight-thousander." What is it about becoming a museum director that appeals to you at this stage of your life?

M: I am only the initiator of the MMM. I need assistants, administrators, and sponsors, too. Putting this idea into practice is more costly—and more difficult—than climbing all fourteen 8000-meter peaks. Once again I am prepared to invest all my energy, time, and means in this project, and once again I might fail, or end up bankrupt. So I am again operating in the area between self-injury and possible self-destruction, and trying to avoid both.

So, yes, in that sense this is my fifteenth eight-thousander. And I've got my best friends helping me, doing the best they can. I never wonder what the final outcome will be; all I want to do is turn a vision into a reality. And because there is so much enthusiasm being channeled into it—from friends, artists, architects, collectors, assistants, and advisers—something unique will be created that will eventually support itself. I never wanted any more than that.

H: Do you find it more satisfying these days to engage with works of mountain art than with mountains?

M: I still climb now and then. Every summer I climb a couple of dozen peaks, and I go away on expeditions once a year. But I am mentally occupied with the mountains every day of my life. Mountain art, mountain people, the history of rock climbing, and the religious aspect of the mountains are the themes that interest me today.

It's exciting to imagine meeting George Leigh Mallory on the summit of Mount Everest or bivouacking with Walter Bonatti at 8100 meters on K2. And I can imagine it better because I've been there myself, in mortal danger, sick with fear and, finally, happy to have survived.

Alpine history as told by couch potatoes is not only boring and sterile, it is false, too, because you don't get a taste of how it feels to be beyond all things with no way back. My concerns are the emotion that is tangible for brief moments between heaven and earth, the perspective we gain when we come back down, the sublime feelings that we have no language to describe.

I am very grateful for the climbing years, the decade on the eight-thousanders, and the months in the deserts, which remain in my memory and can be accessed whenever I like. What is more important, however, is the here and now, the friend who dreams of climbing mountains, the artist who distills the mountains to make them appear even more sublime.

H: Why is this kind of experience more important?

M: My approach today, an experience that I would have dismissed with a smile thirty years ago, is what makes me curious, keeps me awake, and helps me maintain my zest for life. Maybe it's just the opportunity to be creative that drives me or the instinctive knowledge that I need different areas of experience as I grow older.

H: If you compare the museum project to your Antarctica expedition, for example . . .

M: I can only smile. And my rock climbing is not comparable with the high-altitude mountaineering I did either. But one thing is for sure—I have always been obsessive about what I do. Walking across Antarctica was like walking across a foreign planet. And now I am taking on another challenge, a self-imposed task, an objective. This time, however, I won't be going to a summit or a pole, but I might achieve a total work of art that explains the nature of mankind.

If I hadn't experienced those early climbs on the Geisler, the tragedy on Nanga Parbat and its consequences, the near-fatal disaster at the North Pole, and the Gobi Desert trip, I wouldn't know anything—at least not enough to fill a structure like the MMM.

CHAPTER VII
TYING UP LIFE'S
LOOSE ENDS

2009–2014

I can sleep enough when I am dead.
Let me now see the sun and feast on the light.
For I will be hungry for both when I am dead.

—Raoul Schrott, *Gilgamesh*

EPILOGUE

H: Herr Messner, what do you regard as your greatest achievement?

M: Having survived.

H: Which of the defeats do you find hardest to bear?

M: The death of my brother on Nanga Parbat in 1970.

H: There is an old boxing adage that backslapping does more damage in the long run than the punches you take from the front. What did you learn from this defeat?

M: I came to the conclusion that life is limited and only worth something if you exploit its full potential, if you savor it to the fullest. After that tragedy, which nearly killed me, I lived life much more intensely.

H: So the death of your brother did not inhibit you—it freed you?

M: It showed me my limits and my mortality, and challenged me to keep living with double the effort and commitment.

H: What was the most important decision of your life? That fateful descent on Nanga Parbat?

M: No, it was my resolve to live life on my terms, according to my wishes, ideas, and dreams, and not those of my parents, teachers, or brothers. I have never let myself be coerced into a conformist middle-class existence.

H: You have always had a tendency to be a loner. Did the death of your brother strengthen this trait?

M: I would dispute the fact that I have a natural tendency to be a loner. As a young man, I only climbed alone when I didn't have a partner or when the others stayed down below because they thought the weather might turn bad. But I was always more scared of soloing routes than doing them with a climbing partner.

H: But the achievements that revolutionized climbing, and for which you gained the greatest recognition, were your solo ascents of Nanga Parbat and Everest. Where others struggled for months with tons of equipment and dozens of helpers, you carried only a small rucksack and climbed the mountain in a few days on your own.

M: The 1978 Nanga Parbat expedition was the next logical step in the process of reduction. The idea was to do without all that equipment—radios, satellite phones—and ultimately to do without a partner as well. It was a chain of experiences that allowed me to take that last big step. Soloing Everest was about me wanting to test myself. I was looking for the answer to the question of whether I'd be able to cope up there on my own.

H: What do you see as the main problem on solo ascents?

M: The technical difficulties are no greater than they are when there are two of you. But having no one to talk to, no one to share the fear, the joy, and the doubts—that becomes a huge psychological burden.

H: To borrow a phrase from Franklin D. Roosevelt, what you feared most was fear itself?

M: Not during the day. I generally feel pretty good during the day and enjoy being on my own. I even talk to myself as I'm going. It's in the evening that the problems start. It's not that I'm scared of being attacked by a snow leopard during the night or anything; it's just that things get more dangerous when it's dark. When it's dark and you're doing nothing, it's very, very difficult to cope with being on your own.

H: Surely fear of the dark is something that only small children have?

M: We humans need warmth, light, and a feeling of security to survive. When one of these elements is missing, fear creeps in. We all have an aversion to being alone at night.

In 1969 I soloed the hardest rock and ice routes in the Alps: the north face of the Droites and the Philipp-Flamm on the Civetta, both around 1000 meters long. At the time, the north face of the Droites had never been climbed without a fall. I didn't dare start up those routes at two o'clock in the morning, which would have given me the whole day to do them. I didn't leave the hut until it was daylight. In the early morning light, my fears evaporated.

H: There are no cozy huts on Nanga Parbat, though. How did you cope with the nights in the tent?

M: It was hard work. I had to learn to cope with being completely on my own without going crazy.

H: What does the night do to you?

M: The dangers are magnified, and my abilities are diminished. I think it's an instinctive thing. At night it's hard to react to dangers. A hundred thousand years ago we stayed awake when there was danger nearby, keeping watch. Survival was easier the more of you there were, so we banded together in groups or tribes. These primeval instincts are still ingrained in us. I don't have a problem admitting it. I try to come to terms with my instincts.

H: So you prefer to go solo, and if others are required then, that's purely because of ego and self-preservation?

M: I've never tried to hide the fact that I'm an egoist. I've always been honest about it. Every human being is an egoist, and the more we stand with our backs to the wall, the more egoistic we become. We wouldn't survive otherwise. This is not a positive or a negative thing; it's just a fact. To dispute it would be disingenuous. Having said that, I think I've probably done more for the community than many self-proclaimed altruists.

H: That depends how you define it. For many altruists, the egoism lies in helping others. Your egoism is predominantly about helping yourself.

M: And others. How many people have I helped down the mountain, supported, rescued? My approach has allowed me to live my life the way I wanted to live it.

My father wanted to make me into a good citizen; I was supposed to become a poultry farmer. Herrligkoffer didn't even want me to tell anyone what really happened on Nanga Parbat. In the '70s, the South Tyrol bureaucrats tried to make it impossible for me to go on my expeditions because I'd come out with a statement that was not politically correct. Now I'm trying to build my mountain museums, and once again I'm having to overcome huge opposition.

I could claim that all my projects are altruistic, but I don't do that. All I'm saying is this: is it really a sacrilege to do the best you can, to express yourself, to do what I do?

H: In spite of the opposition, you have always won in the end and gotten what you wanted. Today you are the most famous living mountaineer in the world; your books have sold millions of copies. Yet your ego still seems to be unsatisfied. Why is that?

M: Unfortunately—or maybe fortunately—I am very aggressive by nature. A friend of mine who works as a doctor in Zurich once told me that he didn't know anyone else capable of getting so aggressive when faced with a life-threatening situation. I'm like a wild animal. My eyes get really wide, and my body produces an incredible amount of adrenaline. I can feel the energy, the courage, the rage inside me. It's a survival thing, an instinctive reaction. Without it, I—and many of my partners on those extreme climbs—would now be dead.

H: Fine, but what kept you alive in the death zone is not necessarily a recipe for daily life in civilized society, is it?

M: I've often used the same approach when confronting my opponents—those who wished to vilify, oppress, or ostracize me—as I did when confronting danger in the mountains and the wilderness. And achieved my objectives. Maybe I've carried the behavior pattern through into old age because it's worked for me. We humans learn through trial and error.

H: Do you consciously use these outbursts of anger?

M: I can playact, yes. I had another outburst of anger only yesterday, as a matter of fact, and all because of a few bureaucrats who were making things difficult for my organic farm. It was so bad they could have prosecuted me for it. When people mess with me, I can quite easily explode, and there's no guarantee what might happen then. Generally, it's me that suffers as a result.

H: Have you ever thought about taking an anger-management course? It's supposed to have helped Naomi Campbell, and she—apparently—beat her assistant with a cell phone when she was unhappy about something or other.

M: I can control myself. But I won't put up with people messing with me. My time is too precious for that. And as for getting wise in my old age, well, let's just say I haven't got there yet.

H: So the anger will continue, even though it can sometimes get ugly?
M: If I've got an idea and it's legal and it doesn't harm anyone, why should I allow a bureaucrat, an opponent, a wheeler-dealer, or a journalist to stop me from doing it?

H: So who on earth is allowed to criticize you?
M: Anyone, if it makes them feel better. I've had to put up with more criticism than the rest of the climbers put together, much of it unfounded. Now and then I'd like the right to reply, that's all.

H: Another Immoderate statement that's unlikely to win you many new friends.
M: Should I lie or curry favor just to make myself popular? I don't react badly to criticism, and I don't dwell on it. I react when people try to prevent me from putting my ideas and plans into practice, some of which will ultimately be of benefit to everyone.

H: Your first wife, Uschi Demeter, once said this about you: "I don't know anyone who wants so much to be loved yet can do so little about it."
M: I'm not going to pretend to be something I'm not—to smile sweetly all the time—just to get recognition and love. I won't play the whore for anyone, least of all for journalists. I can't do that. I want to be loved, yes, but the way I am and not the way people would like me to be. I have to be allowed to be myself.

H: With all due respect, you owe your success not only to the fact that you have climbed some big mountains but also to the fact that you were able to tell people about it. Many of your expeditions were financed in part by magazines, and over the years many journalists have raised your fame. Without the modern media, the phenomenon of Reinhold Messner would never have existed.
M: That's anyone's guess really. I've always gone on the stage and told my stories. I am an adventurer and a bard; I come home and I tell my story. Above all, I am a self-made man. The fact that the media seize hold of my

stories has helped me, for sure. I've never complained about that. But they don't have to distort my stories in order to sell them over and over again.

H: The media have profited from you, and you from the media.
M: Yes, there's always give-and-take. We struck deals over the financing of expeditions. A good story always has a financial value, too. Why should I give it away for nothing? However, I've never scrambled to get interviews or publicity. When I go on my trips, I need to be a free agent. Nor have I ever told stories just because they would sell well.

Others have curried favor, but not me. I know how it works, of course. All you have to do is wax lyrical about getting close to God on the summit, about idealism, noble comradeship, and the fact that climbers are better people—about loyalty unto death and all the other stuff that goes to make up the popular view of mountaineering.

My life has given me a different view of the world, and of human nature. It's shown me that up on the summits, people are exactly the same as they are down here. There are a lot of nice people who go climbing: lads and lasses, children and old men. But there are also some bad characters with serious hang-ups trying to compensate for their failed ambitions, people who talk about comradeship in the mountains but don't know where the mountain is or that a willingness to help should be a matter of course.

H: The journalist Wilhelm Bittorf, another great admirer of yours by the way, once wrote that being continually challenged in extreme situations has given you an enviable confidence in life. Normal everyday worries, he concluded, are alien to you. Problems like "Who will pay for my pension?" or "Will my car pass its inspection?" seem not to affect you. Is this diagnosis still applicable?
M: It's still accurate, yes, but I don't think my carefreeness has anything to do with the many extreme situations I've faced in my life. I think it stems from the fact that I've always managed to keep moving my life forward, in spite of all the obstacles that were put in my way, in spite of all my mistakes.

Even when I fell off the castle wall and the doctors told me I'd never walk again, there was no angst, no existential fear. I simply told myself I'd find something else to do, even if I had to do it in a wheelchair. As long as I have my set objectives, I am confident.

H: Do you think that might have something to do with the fact that, at the time of the accident, you were financially and materially secure? After all, you were the owner of a castle and a working farm.

M: Owning things is boring—obligations and responsibilities. It interests me less than creating things. Right now, I'd be prepared to give everything I own to my children and start again from scratch. The farm and the self-sufficiency do give me a good feeling, though.

H: And you won't go hungry.

M: What gives me strength is the feeling of being independent. In effect, I'm really just a dilettante. I've lived, explored, and worked—but only in nonjobs. I've often achieved success against all the predictions. And I've done it by following a very simple pattern of behavior: stick at it and do everything in my power to make it happen.

No reasonable person would invest the time, energy, and money I'm investing in my museums. My financial advisers just shake their heads. The passion I have for what could be viewed as a futile pursuit has made me strong, and it is this that gives me the confidence required to lead a self-determined life.

H: What advice would you give to those who are totally passionate about something yet fail to achieve success?

M: It obviously helps if you have a talent for what you do. It's hard to get enthusiastic about something that doesn't suit you. If I have a lung disease, I'm not going to get very far with mountaineering; I'll never get to the summit of Everest, for example. Having a physical injury can also take away some of the drive and passion. When I lost my toes to frostbite on Nanga Parbat and wasn't able to climb as well, my passion for my first love, climbing, was lost. It began to fade. Fortunately, I discovered something else, which replaced the old passion: high-altitude mountaineering.

In later life, too, the fact that I was able to change direction again and again—and always at the right time—was a distinct advantage. Throughout my life I've usually realized when it was time to say, "That's enough of that; I need something new."

H: You once said that you only find things exciting when they are new. When you look back at your achievements, do you find them boring?

M: I even get bored when strangers pat me on the shoulder and congratulate me on my successes. The only time the way I feel about what I've achieved changes is when I'm onstage talking about it. That's when it all comes back to me and I relive it all—the pack ice, the storms, the rock and the ice.

I am the storyteller and the protagonist in one and the same person. I'm part of the story, yet I'm also detached from it. My skills as a storyteller are based on the ability to immerse myself in past situations and different characters.

H: Do you perceive this constant drivenness, this inability to look back at things in a calm and relaxed manner, as a shortcoming?

M: No, I see it as good fortune. I couldn't cope with life any other way. And I can always get wise when I'm older. A last hope.

H: You also said, "When I no longer have any dreams, I'll kill myself." Do you need this kind of emotionalism to drive yourself onward?

M: I don't think you necessarily do everything you say you'll do. Or should that be the case? No, suicide is not my thing. I hope I will always have dreams.

H: You don't find the pathos in that statement embarrassing?

M: No.

H: Do you have any character traits that you'd like to change?

M: My occasional outbursts of anger and my impatience. If I gave myself a bit more time, a lot of problems would solve themselves.

H: How long can you put up with waiting in a restaurant for a plate of spaghetti?

M: I can wait, and I can put up with a lot. What annoys me are things like bureaucracy, envy, and exclusion, marginalization. I have to tackle problems like that immediately if I want to drive my projects forward.

Now and then I'll let something lie and six months later realize that the matter has happily been resolved. In many cases I should just force myself to leave things alone and let them run their course. Not always, but

in many cases. I'm in the last third of my life now and shouldn't feel con-
strained by anything, apart from my own infirmities.

H: Is friendship important to you?
M: Yes, very.

H: How many friends do you have?
M: Not many. And I'm proud of the fact that I don't have many.

H: More than a handful?
M: Barely.

H: Why?
M: You need time for friends, and you also need to be on the same wavelength. I met people in politics who I got along well with, but I wouldn't
call them friends as such. Friendship requires trust, empathy, and accep-
tance. I don't want to feel that I have to wear a disguise in front of my
friends; otherwise, the friendship is worth nothing. Friendship means being
accepted by a person for the way you are, and vice versa.

**H: That sounds a little passive. Doesn't friendship also mean making an effort
to meet halfway, to compromise?**
M: That, too, yes. But I don't have to keep changing the way I am. If someone
has a habit of lying but he's my friend, I have to accept it.

H: Difficult to trust that person, though.
M: Yes, but I know my friends. And friends have their problems, too. You
are allowed to show your friends the real you. Who else is there?

**H: Which of your expedition partners—Habeler, Fuchs, Kammerlander—would
you still describe as a friend?**
M: We meet up from time to time. But we all live in our own separate worlds
nowadays. We shared experiences. We formed partnerships of convenience.
But the relationships are not the same as they were, and we all tried our luck
with new partners.

H: Fuchs?

M: His manager tried everything he could to force us apart. He succeeded, and Arved let it happen. But hey, what the hell—he taught me how to navigate, and I thank him for that.

H: Habeler?

M: Of the three, he's the one I see most often. We've both grown older. Peter has a successful career.

H: Kammerlander?

M: Not my cup of tea. He was an excellent alpinist of course, a good guy to have with you on bold, ambitious ventures. But I can't ever imagine myself chatting to him about my museum and such. He's been successful, he's played the game and had a few books published, and now he follows the trend of "piste alpinism," as I call it. That's not my world.

Nowadays, my world is less defined by action and more by my commitment to the collective mountain legacy. With Hans it's all about publicity, whereas I'm more concerned about the background. Hans was the strongest climbing partner I've ever had. But a true friend is there for you in difficult times as well.

H: Which of the people you have climbed with would you still describe as friends?
M: None of those three. With them it was more a case of partnerships of convenience and friendly relationships rather than real friendship. One friendship that has lasted for forty years is my friendship with Sepp Mayerl. Bulle Oelz is also a friend. And whatever happens, there's always Nairz, Heim, Hanny . . . I'm still close to my childhood pals from Villnöss, but I've lost touch with others. My friendship with Hanspeter Eisendle has grown over the years. We hardly ever go climbing together these days, but we do share other interests.

A big trip with friends to climb the south face of Lhotse is no longer possible—all my pals from earlier are too old for that now, the same as me. For objectives like that you need the skills, for sure, but you also need total trust in your partners. When that basis is no longer there because trips like that are no longer possible, you still have the good memories, but it's hard to keep the friendships alive unless you find new things to do together.

H: Uschi Demeter once described you as a "predator." Do you regard that as a compliment?

M: I can certainly motivate people and get them involved in my projects. I would also say that in the majority of cases it can be established that my partners at the time were so involved with our common objective, right from the start, and had internalized it so completely that they were of the opinion that it was their own idea they were pursuing as well. That's the only way to cope with extreme situations when you're pushing the limits.

None of my climbing partners would have been prepared to go to Everest with me and assume the role of porter when there was a chance they might die. If it isn't something you really want to do yourself, at some stage you'll end up wondering why on earth you should keep going with the idiot who thinks he can climb the highest mountain in the world without an oxygen mask.

H: With hindsight, does it annoy you that on many of the trips where you were "pushing the limits" you had to do so with a partner?

M: No, not at all. I learned a lot from them. And having partners made a lot of things possible that I wouldn't have dared to try on my own. When you approach the limit of what is and isn't achievable, the line between self-injury and self-destruction, between life and death, you need confidence, skill— and a good partner. I need my partner first and foremost for psychological support, which is why I've always opted for people who could do things I couldn't do.

H: Who is the greatest love of your life?

M: My wife, Sabine. I had a strong bond with Uschi, too, and I shed a few tears when we split up. Not anymore, though.

H: You will be seventy years old this year. How old do you feel?

M: Ageless. When you are seventy, the chances of dying in a dangerous situation you've chosen to put yourself in are not quite as great. That means I've now got a really good chance of growing old. The only question is how. The curiosity remains, and that's a good feeling. Above all, I now have the peace and calm I need to create my museums and to foster a few friendships.

H: What do you look forward to in the future?

M: The harvest is in, and the barn is full. I just have to do what needs to be done—get rid of the rubbish and keep the place clean and tidy. I look forward to tomorrow.

H: What is likely to remain important to you?

M: There will always be something: the two-hundred-year anniversary of the first ascent of the Ortler; the Gobi Desert expedition, a trip that once again took me right to my limit; learning to come to terms with growing old; some kind of political swan song, perhaps. I also plan to make some trips to places that have been contaminated by industry, in the former Soviet Union, in China, and on the shores of the Arctic Ocean—places where no one lives anymore but the oil tap is dripping.

There are enough challenges to keep me busy in my old age—lower mountains and smaller deserts—and plenty of fun to be had, just like it was when I was twelve or thirteen years old. I'd like to make another film, too, and write a few good stories. Maybe take an idea and start again right from the start . . . my seventh life. There will be no time for idleness, that's for sure.

Through my failures I have learned how to live—and the more I failed, the more I learned. In my search for the limits, I have failed more than most, and it is this that has made me successful, over and over again.

H: Which literary figure do you most admire?

M: There isn't one. I have always looked for my role models in the real world.

H: Which historical figure would you describe as a role model?

M: Paul Preuss, the inventor and philosopher of the art of free climbing. His personality really comes through in his writing.

H: Yet this is a man you expose in one of your museums by exhibiting the hammer which he apparently used to hammer pitons into the rock in secret. Why do you have to try to get the better of your only hero in this way?

M: The exhibit in the Alpine Curiosa museum in Sulden is a declaration of love to a dead friend, not a demolition of his character. It's that hammer that makes Paul Preuss human.

H: The last of your six museums is due to open this year.

M: Yes, the MMM Corones. It's dedicated to traditional alpinism. The Plan de Corones is the Ladin name for the Kronplatz, the mountain above Bruneck where the museum is situated. It's a very popular ski resort, so it's really busy in winter, but in summer there's a lot less going on. The architect Zaha Hadid designed the building.

Classic alpinism is in decline. Nowadays it only accounts for 10 percent of activities in the mountains; the other 90 percent is made up of tourism and other sports. This 10 percent is very dear to my heart, and I'd like to guarantee it some kind of continuity.

H: Does your own story also feature in the new museum?

M: My life is closely linked with traditional alpinism, so I've started with my own development as a climber—where I come from, and who my predecessors were. The museum provides a brief history of mountaineering—from Albert Mummery on Nanga Parbat to the present day—all told through individual stories about specific mountains and specific faces. The whole thing is illustrated with paintings, black-and-white and color photographs, and moving images, too.

H: The MMM Corones is the culmination of a complex and expensive project. Fifteen years of work, at a cost of 30 million euros [about US$40 million].

M: It wasn't all my money. I only supplied two buildings, plus all the fixtures and fittings, and sourced the exhibits. My job is to curate the six museums and to run them without subsidies.

H: Why did you take on the burden of responsibility of managing these museums for thirty years without payment?

M: The thirty years only applies to the museum at Sigmundskron. In Bruneck it's twenty years; at Monte Rite and the Kronplatz there's no obligation.

H: Nevertheless, you will still have to maintain the exhibitions without subsidy for several decades—in one of the museums at least. How much will you have to invest yourself?

M: It's certainly been a lot. I can't tell you the exact amount, but for a while my wife was worried we'd lose everything we owned. But fortunately, it's now looking like the project will finance itself in the foreseeable future. I'm not complaining. The problem I have is finding someone to continue my work. My contractual obligation still has another seventeen years to run.

H: Is there anyone in your family who could take over from you?
M: My daughter Magdalena has the ability. She is not a climber, so she doesn't have the in-depth knowledge of climbing that would make the job easier to do. A knowledge of art is more important, though.

H: You once said that money means nothing to you. Do these museums now mean anything to you?
M: The money isn't important, and it's not necessarily about the end result either—what is important to me is being able to continue the creative process. That is what really interests me. The money is necessary only to fund that process. One thing is certain: I will carry on being creative in the future. And I can only play the game if I have the money for the stakes.

H: *Homo Messner, homo ludens?*
M: Well, yes—but not to the point of overindulgence like some people I could mention. I get a kick from finding exhibits and developing ideas. As far as the museum project is concerned, I get a lot of pleasure from finding things and bargaining for them. I've often spent months haggling over a single piece. It's like going to market and buying a cow. It's become a passion of mine, using every means at my disposal, including my know-how, to build something unique. No bank in the world would have lent me the money for my museums.

H: Why not? Your museums contain valuable items that could be used as security.
M: The banks all assume that these museums only work because I am behind them.

H: So you only invest the money that you have already earned?
M: Yes, and I want to use the money to do things I'm passionate about—in this case, to develop the museums. In the past I used the money I made to go on expeditions. As soon as I'd saved enough, I'd be off again—to the

Himalayas, Antarctica, the Gobi Desert. Then there was Juval to pay for. I made money as a freelance writer, selling garbage. That's how I created the free space to do what I wanted to do.

H: You describe your writing as garbage?

M: To some extent, yes, although I must say that the money I've made from writing has been the hardest earned. If you work out the hourly rate, I get paid much more for everything else I do. You get more for chopping wood than you do for writing.

H: You'd have to chop really quickly, though.

M: It's true. Getting wood from the forest, chopping it into logs, and then selling them brings in more money than writing. Chopping wood is also nice, physical work, unlike writing, where you sit and work and then have to wait for ages until you see your story in print.

Newspapers no longer pay anything for your stories either; they expect you to do them for nothing. I don't see why I should write a piece for *Süddeutsche* only to be told that of course they will publish it—but they won't pay me a fee. I won't do that. It's not right. Every job has its price.

H: So you would rather chop wood?

M: I could also keep geese or slaughter pigs. There are lots of options. But writing without payment, no.

H: Your farm is now leased out. It would appear that being a farmer was not much fun for you either.

M: It's important that people actually understand what it is that I want—a fair farming system. That only works if I produce the food on my own farm and sell it ready to eat, straight to the plate, as it were. If I have to take my milk to market—maybe sell it to a dairy factory, where it's made into cheese or yogurt—I get such a low price for it that I might as well not bother.

H: Wicked tongues in South Tyrol claim that you have created a "Messner mausoleum" for yourself with your museums. Does this kind of malicious gossip bother you?

M: That's only because they think I'm narrow-minded. I would be pretty stupid if I opened a museum that was all about me. It would last a year, maybe five at most. I have built museums that showcase the relationship between man and the mountain.

H: One of the exhibits in the MMM Firmian features a boot that belonged to your brother Günther, from the fateful Nanga Parbat expedition. When and where was that found?
M: In 2005, three and a half kilometers below the place on the glacier where Günther was buried in the avalanche. The boot is a Lowa Triplex, and that particular model was made especially for us for our expedition— black leather outer boot, leather inner boot, a removable felt lining, and red laces.

We managed to get the foot out of the boot, as it was relatively well preserved. The results of the DNA analysis carried out on the foot at the University of Innsbruck were clearer and more exact than the test results on the bone that Eisendle found in 2000. In 2003 they said that the DNA match had the accuracy of a paternity test. We now had the kind of accuracy you get maybe once in a hundred years.

H: Previously, you just had bones, but your brother's boot was more concrete evidence, I would imagine. Did finding the boot provide you with a kind of emotional closure?
M: It meant we were finally able to hold a funeral for my brother. After the first bone was found, I asked the locals to keep on searching in the same place, as I was sure that more of Günther's remains would come to light. They found a lot—his backbone, his pelvis, and several hundred bone fragments—spread over an area of about 3 meters by 4 meters. After all, a glacier is a flowing mass of ice, so the avalanche snow that originally covered him had melted and been carried down toward the valley over many years, leaving the bones exposed.

We cremated Günther's remains there in Pakistan and built a memorial on a rock at the spot where the base camp for the Diamir Face is now situated. In 2007 my family went there together—there were twenty-four of us in all—but the memorial plaque had unfortunately been removed.

H: Back in 2003, when the first bone went to the Institute of Forensic Medicine in Innsbruck for DNA testing, you were very upset and hurt, as you were still dealing with that libel case, with people saying that you had sacrificed your brother to your own ambitions. How do you feel about it now, after all the findings and the funeral?

M: The anger and indignation are behind me now. I will no longer defend myself against the accusations. I have nothing more to say on the subject. Doing so just fuels the conspiracy theories. The smear campaign made good business sense for my detractors. They sold a few books, and for a while people believed them. No one wanted to hear the facts. It is a fact that my brother was buried beneath an avalanche. The responsibility for that rests firmly with me; no one else has to take the blame. But having to prove that the others had lied really was too much.

Anyway, it's over now. If someone says my museums are rubbish — fine. If someone comes along and says "Messner never climbed the seventh grade"—that's fine, too. I couldn't care less. It's all just self-righteousness on the part of the doubters. As far as I'm concerned, they can make things up as they see fit. I'm staying out of it. There's just one thing I ask: please let me live out the last years of my life as I see fit.

H: The director Josef Vilsmaier made a film of the Nanga Parbat story for the big screen. This brought the dreadful tragedy of that ill-fated descent to the attention of a wider audience. How closely did you work together with Vilsmaier on the film?

M: I was on location and offered my advice. It is a classic mountain film. I think it's okay; well, not bad anyway. The two Messner brothers are not as believable as Herrligkoffer, though. [Karl] Markovics plays Herrligkoffer so well that he becomes a real personality. Herrligkoffer was never really such a strong character.

H: You and your brother feature very prominently, of course. Vilsmaier was well advised not to have big stars playing the two brothers.

M: I agree. They had to be young and low-key. However, one of the big mistakes was that Vilsmaier never went to the Diamir Valley himself.

H: In spite of that, the film still contains some powerful moments. You and your brother having to bivouac just below the summit without a tent at minus forty degrees is something the audience will never forget.

M: That's one of the best scenes. In the bit where I shake my brother and ask him, "Are you still alive?" and our clothing is frozen stiff, you can actually see, and feel, how cold it is. It's impossible not to empathize with the two men.

H: Then there was the documentary film *Messner*, which features some of your brothers.

M: One of the reasons that film is so good is because my brothers have brains and can express themselves well. They all have doctorates—even the ones who don't appear in the film—unlike me. I wasn't there when they were filming with my brothers; they had enough to deal with, and I didn't want to bother them. But they provided a vivid psychological portrait of me as a person.

H: Your brother Hansjörg is a successful psychoanalyst in London. He describes you as a man who is driven by a compulsion to repeat.

M: Hansjörg says some good things in that film, although I don't always take what he says very seriously. It's a bit tongue-in-cheek sometimes.

H: Hansjörg was a bit of an adventurer himself as a young man. There is the story of how he went missing in India and your mother kept asking you as you were about to set off for the Himalayas, "If you see Hansjörg over there, will you please bring him back home with you?"

M: I didn't find him in India. Later, in 1977, I was walking through the streets of Kathmandu when I bumped into him. I asked him what he was up to, and he answered, "I'm just wandering around." I invited him on the expedition, and he accepted. He was in good shape, not as battle-weary as I'd expected. When we got to base camp, I told him that our mother was worried about him and that he should probably come home with me. He said he would come home but that he didn't want to fly back with me; he wanted to travel overland. He asked me to give him what the plane ticket would have cost in cash.

Two or three months later, he was back home in South Tyrol, having hitchhiked most of the way. Then he decided he wanted to finish his high school diploma, so my brothers and I helped him. He ended up working for Janov in London before he and his partner took over the clinic from him. He now treats people all over the world, from the United States to Australia.

H: Have you ever had psychotherapy?
M: No, I don't believe in that kind of cure. I don't think psychoanalysis is nonsense necessarily; I just don't see the point of it. An analyst can find out where the problems lie, but a witch from the jungle can do that as well. If I'm not allowed to live my own life, no psychologist is ever going to be able to help me. The real problem is that most people don't live their lives, they just get by somehow. I believe in genetics, not the advice of magicians!

H: But surely the purpose of psychotherapy is to create harmony within the person?
M: I can get that balance when I go climbing from time to time with my son, Simon, and I realize that I'm not as good at it as I used to be. Simon suffered from vertigo as a boy and only started climbing later in life. He's very good on the rock now. He's an elegant climber, but he's clever enough to realize that he doesn't want to pursue a career as a mountaineer. With me as his father, there would be too much pressure on him.

H: In 2009 you married Sabine, who had been your partner since 1985. Why did you decide to make it official? After all, the relationship had worked well for twenty-four years as it was.
M: She never wanted to get married, but the kids started pestering us to do it. They said we'd been together so long that it was silly not to, and that we were just being pompous. So we thought, "All right then, we'll get married!" Our little daughter wanted a horse-drawn carriage and all the trimmings, but we didn't go for all that bourgeois stuff, naturally. We managed to avoid all the media stress by bringing the whole thing forward one day.

H: A small wedding then?

M: A very small, formal act. No ties, no suits, no magic spells.

H: Is living together any different now?

M: Not at all.

H: And you are still allowed to do your own things. However, it seems as if what used to be regarded as adventure has now acquired some pretty bizarre characteristics. The highest mountain in the world, Mount Everest, where this year you spent three weeks, has become a caricature of bravery. Why do you even bother with this kind of luxury mass tourism?

M: I was there as an observer. On Mount Everest, clients pay up to US$80,000 to be guided from base camp to the summit—up a pre-prepared piste! Five hundred Sherpas spend a month preparing the route, with ladders on every steep step, bridges across every crevasse, and fixed ropes on each side—so it's impossible to lose the way. For most people up there, it's a holiday; they have saved for it. The guides are like kindergarten teachers: "Right, we're setting off now; drink your tea now; there will be soup waiting for you at every camp; time to go to bed now; I'll wake you at four; put this jacket on; don't forget your boots; wear these crampons—and off we go."

H: It seems like the biggest danger nowadays is being trampled underfoot by all the other aspiring summiteers up there.

M: They could certainly use a traffic policeman on the Hillary Step, or maybe a solar-powered traffic light up there. At base camp in 2013 there were four helicopter landing pads. At times, with helicopters coming and going, it was louder than Munich Airport. One woman had herself flown back down from Camp 2. The pilot told her, "You do know that it costs twenty thousand dollars?" Her reply: "Twenty thousand or eighty thousand, it doesn't matter. Just get me down." In a few years, all the eight-thousanders will have prepared pistes; I am convinced of that.

H: Could you still get to the top of Everest?

M: No problem. I'd have to use bottled oxygen nowadays and the prepared trail, but I could do it. I would find it dreadfully embarrassing to do it like that, though.

H You'll be seventy this year. Do you notice your age when you are climbing?

M: I don't acclimatize as well as I used to. My circulation isn't as good. On our 2014 trip to Tibet it was minus twenty-five and I had cold feet all day. My blood pressure used to be extremely low; it isn't anymore. The main thing is, I'm not as agile as I was when I was fifty; I stumble a bit more. My son, Simon, is much more agile and moves more smoothly on the rock.

From thirty onward your climbing starts to go downhill. The first thing to go is the explosive strength, then the agility, then, at about sixty, the endurance. Finally your capacity for suffering decreases to the extent that one small pain is enough to stop you climbing well, as it interrupts the flow.

H: Angela Merkel, the chancellor of Germany, doesn't climb, but you have been out with her in the mountains of South Tyrol. How would you rate her endurance?

M: Well, firstly, she is very modest. And it's not pretense. As a young girl growing up in East Germany, she dreamed of going to the Alps but was not allowed to. She doesn't have the time to train these days, but she is tough and persistent. She just keeps going, meter after meter. A thousand meters of height gain in two and a half hours is not bad going!

She looks for peace and tranquility in the mountains. The provincial president of South Tyrol once asked her if he could join her on a hike. She told him no, that she didn't want to see any politicians up there. That woman is full of integrity, 100 percent.

H: As a former Green politician—would a "black-green" coalition be your preferred political alliance in Germany?

M: Absolutely. A coalition between Angela Merkel's Christian Democratic Union/Christian Social Union bloc and the Greens would make Germany a parliamentary role model in world politics again. The Greens have a few strong politicians: the minister-president of Baden-Württemberg, Winfried Kretschmann; the mayor of Stuttgart, Fritz Kuhn.

H: Both political realists and both in favor of a black-green coalition.

M: But Herr Trittin screwed it up. He is arrogant and lacking in authenticity. And he proved with his tax plans that he is far removed from the practicalities of politics.

H: Last year you had a break-in at Juval. You caught the intruder. Did you call the police?

M: No, I didn't call the police. I'd left the door open for my daughter Magdalena, and I was sitting reading, when just after midnight I heard footsteps. "Good," I thought. "We can chat for a bit before bed." Then I heard a man coughing. I went into the living room and saw a shadow behind a pillar. There was a man with a hat on, and he was bigger than me. I thought about getting a knife from the kitchen (I don't have a gun in the house) but decided against it.

I felt an uncontrollable fury, the tremendous aggression I can summon when I feel threatened—my best weapon. With a roar like a war cry, I just went for the intruder and grabbed both of his hands. I held them so tightly that I tore a tendon in my left thumb. I dragged him to the bridge, where there's a thirty-meter drop, and told him, "You try that again and I'll throw you down there." He was so scared he was shaking all over. "Please, no," he said, whimpering by now. I let him go. He disappeared, and he's never been back.

H: For many years, Death was one of your closest companions, challenging you to achieve peak performance. After which, as you yourself have said, you felt reborn. As you approach the age of seventy, does that change your relationship with your old companion, with Death?

M: He's just around the corner and closing fast. The time I have left is decreasing. All I really want to do now is put a few ideas into practice. I don't want to waste any more time with self-righteous critics or people wandering around my castle at night. I would like the children to take over from me. I'm not an administrator, nor do I enjoy patching up old building sites.

For me, "success" is not measured at the end of a life. A successful life is what you have when you are doing things. There are moments when I overcome difficulties, and that's when I feel strong and fulfilled. My success, my life, was nothing more than turning ideas into reality.

H: You already speak about your life in the past tense?

M: Yes. I don't really have to do anything more now. But I will continue to fill my time implementing ideas. I am incapable of doing anything else.

Architectural rendering of the sixth Messner Mountain Museum, the MMM Corones designed by Zaha Hadid

H: Do you ever experience feelings of gratitude?

M: Of course. I've had a fantastic time. The generation of climbers before me got to go on maybe one or two expeditions in their whole life. We could do virtually anything we wanted to do. We could live out our wildest dreams. We were even able to finance the trips ourselves. We didn't have to carry the Austrian, German, or Italian flag up the mountain with us just because we were being funded by the state or some national association. We got the money we needed from private enterprise, which meant that we developed an objective relationship with what we did, and with our successes. There were opportunities for all of us.

It was a crazy time. And I was lucky enough to survive it; lucky that I didn't make that one big mistake; lucky that, from 1972 onward, I was able to follow my dreams; lucky to share my life with a few strong women; lucky that I was able to get back down the mountain in time before things got too desperate. Recently, when I was leaving the auditorium at the end of one of my lectures, someone shouted out, "You've been lucky!" That's right, I thought, I've had more luck than any man deserves.

H: Let us assume for a moment that we could turn back time and you were in your midtwenties again. Could you compete with the climbing stars of the present, men like David Lama and Alexander Huber?

M: It's harder to become a top climber these days; the bar is set much higher than it was in my day. The standards that people like David Lama and Hansjörg Auer are operating at are so high that I would never be able to achieve them. I was never that brilliant.

Climbing has become a global phenomenon. The top professional climbers do nothing else apart from climb. They have one or two sponsorship deals, and everything is hyped up so they can finance their routes. I had to give lectures and write books in order to finance my expeditions.

Nowadays, 90 percent of climbers climb indoors. Most of them never climb outside at all. Indoor climbing is just pure sport; it has nothing to do with adventure. But the few top athletes who make it out of the climbing gym and into the mountains—[Alex] Honnold, for example—climb several grades harder than we ever did.

In my time I was regarded as a suicide candidate; they said I wouldn't live very long, that if I kept it up I'd be dead within a few weeks. But the stuff that Alexander Huber is climbing now—free-soloing the Hasse-Brandler Direttissima on the Cima Grande, for example—I would never have dared to attempt. And then there's Honnold, [Chris] Sharma, and [Adam] Ondra—unbelievable what those boys can do!

H: Although that kind of rock gymnastics has very little to do with classic alpinism, surely?

M: Yes and no. Classic alpinism has decreased in popularity. I still go out with Simon and do long middle-grade routes in the Dolomites—big 1000-meter faces, some of them. In the past they would have had fifty or more ascents over the summer. When the weather was good, there was always at least one party climbing on the Civetta every day. These days, if I spent a month there I'd probably only see one team climbing. People don't go climbing in the mountains anymore; they go to indoor walls with plastic holds.

H: Is that a bad thing?

M: Not really. You don't always find happiness in the mountains. It just happens. And then, when you realize that you were happy, it is too late.

It's got a lot to do with what the Americans call "flow," the concept of being in the zone. It is only with the benefit of hindsight that I have found out that I was happiest when I was doing my own thing, and it was often unquantifiable.

Incidentally, that is one of the reasons why I still cause offense in Germany. They think I'm egotistical. When I am implementing my museum ideas, I am doing my own thing—and I do it with great enthusiasm. The fact that it might also bring benefits to South Tyrol in terms of local tourism is just a side issue for me. My feeling of satisfaction, my happiness if you like, is not dependent on applause or accolades of any kind but on the fact that I was able to do what I wanted to do—and to see it through. The museum thing proved to me that my way of life would also work outside of climbing and expeditions. Or it has done up to now.

Perhaps the true purpose of life is simply to express ourselves as best we can. Maybe my ability to keep finding new challenges appropriate to my age is part of the happiness, the thing that keeps me young, creative, and full of life. The setbacks and the opposition I've encountered are all part of my happiness. I have grown as a result and am still able to lead a self-determined life.

H: One of the paradoxes of life is the fact that happiness sometimes just befalls us.

M: Good feelings sometimes show up unexpectedly. It might be a nice turn of phrase when I'm writing, sharing a glass of wine with a friend, a walk in the woods, watching my children play, an evening with my wife, or an idea I have. And when I don't feel good, I simply go for a walk.

MILESTONES

1944 Reinhold Messner is born on September 17 in Brixen, South Tyrol.

1949 Messner becomes interested in climbing at the age of five. He has one sister and is the second oldest of eight brothers. He grows up in Villnöss and attends the Geometerschule in Bozen.

1956 Messner climbs his first difficult rock route, the east face of the Kleine Fermeda, with his father.

1960 Günther and Reinhold Messner climb the north face of Sass Rigais.

1963 Messner climbs his first grade VI rock route, the Tissi Route on the First Sella Tower, and his first ice route, the north face of Similaun in the Ötztal Alps.

1964 In early summer, Günther and Reinhold Messner climb a series of hard rock and ice routes: the north face of the Vertain, the north face of the Hochfeiler, and the north face of the Ortler. In July, Reinhold Messner, Paul Kantioler, and Heindl Messner climb the north face of the Furchetta.

1965 Günther and Reinhold Messner make the first ascent of the north face of the Grosse Fermeda. That summer, Reinhold Messner goes on to make further first ascents in the Dolomites: the south face direct on the Neunerspitze in the Fanes region and the northwest pillar on the Odla.

1966 In June, Messner fails the test for his high school diploma. He spends the summer climbing with Heini Holzer on the Civetta, where he makes several first ascents. Although he fails his exams the second time around, his older brother finds him a post as a supply teacher in Eppan. Messner teaches mathematics and science there for a period of nine months.

1967 In February, Messner makes the first winter ascent of the north ridge of Monte Agner and, in March, the first winter ascent of the north face of the Furchetta. He follows these with further second and third ascents in the Dolomites and a new route on the northwest face of the Civetta.

Messner takes the test for his high school diploma as a private candidate and passes.

1968 Messner makes the first winter ascent of the north face of Monte Agner and, with his brother Günther, the first ascent of the Eiger's north pillar. Numerous first ascents follow, including the central pillar on the Heiligkreuzkofel, their hardest climb.

Messner registers as a student of civil engineering at the University of Padua.

1969 Messner participates in an expedition to the Andes. On his return he makes the first solo ascent of the north face of the Droites, a route he failed on with Günther four years before. He follows this with the first solo ascents of the north ridge direct on the Langkofel, the Phillip-Flamm on the northwest face of the Punta Tissi, the South Face Direct on the Marmolada di Rocca, the Soldà Route on the north face of the Langkofel, and the Meran Route on the north face of the Furchetta.

Messner is still registered as a student, but he does not attend any lectures or take any exams after the Andes expedition. In the autumn he receives a phone call from Karl Herrligkoffer, who invites him to join his forthcoming expedition to climb the Rupal Face of Nanga Parbat. Messner abandons his studies in Padua and returns to South Tyrol to train and work part-time as a math and physical education teacher.

1970 On June 27, Günther and Reinhold Messner top the Rupal Face of Nanga Parbat (8125 meters). They are forced to descend the Diamir Face (via the Mummery Rib), thus accomplishing the first traverse of Nanga Parbat. Günther Messner is buried in an avalanche at the foot of the face.

1971 Messner gives up his teaching job. He starts giving lectures and writing books, and works as a mountain travel guide. In the autumn he travels to the Diamir Valley to search for his brother. He is accompanied by his partner, Uschi Demeter. In the same year he establishes two new routes on the Carstensz Pyramid in New Guinea and makes the first ascent of Puncak Sumantri Brojonegora.

1972 Messner makes the first ascent of the south face of Manaslu (8156 meters). During the night the climbers are caught in an unexpected snowstorm. Messner's fellow expedition members Andi Schlick and Franz Jäger fail to return to camp. Their bodies have never been recovered.

Messner climbs Noshaq (7492 meters) in the Hindu Kush.

In summer he marries Uschi Demeter.

1973 In early summer, Messner returns to Nanga Parbat with the intention of making a solo ascent. The attempt fails. Back home, he makes the first ascent of the west pillar of the Marmolada and the west face of the Furchetta.

Messner and Demeter buy and renovate a former rectory on a hill above Saint Magdalena in the Villnöss Valley below the Geisler.

1974 During an expedition to the Andes, Messner climbs a new direct route on the south face of Aconcagua (6959 meters). An attempt on the south face of Makalu (8485 meters) in May ends in failure. In August, Reinhold Messner and Peter Habeler climb the north face of the Eiger in the record time of ten hours.

1975 Messner fails on the south face of Lhotse (8511 meters). A few weeks later, he and Peter Habeler climb Gasherbrum I, also known as Hidden Peak (8068 meters), via the north face in "alpine style," i.e., without high camps, porters, or fixed ropes.

1976 Messner makes the first ascent of the "Wall of the Midnight Sun" on Mount McKinley (6096 meters).

1977 A spring attempt to make the first ascent of the south face of Dhaulagiri I (8167 meters) is unsuccessful. Back in Kathmandu, Messner boards a light aircraft and takes a flight over Mount Everest without an oxygen mask.

Messner's marriage to Uschi Demeter disintegrates.

1978 In spring, Messner makes the first ascent of the "Breach Wall" on Kilimanjaro (5895 meters). On May 8, Messner and Peter Habeler make the first ascent of Mount Everest (8850 meters) without supplementary oxygen. On August 9, Messner makes the first solo ascent of Nanga Parbat via the Diamir Face. This is also the first solo ascent of an 8000-meter peak.

1979 In February, Messner travels to the Sahara and climbs several new routes in the Hoggar Mountains. In summer, together with Michl Dacher, he makes an alpine-style ascent of the Abruzzi Ridge on K2 (8611 meters).

1980 On August 20, Messner makes the first solo ascent of Mount Everest.

1981 Messner travels with an expedition group to Tibet. He climbs Shishapangma (8012 meters) and makes the first ascent of the north face of the central summit of Chamlang (7317 meters). An autumn attempt on Makalu is unsuccessful. Messner's partner, the Canadian Nena Holguin, gives birth to their daughter, Láyla, in Kathmandu.

1982 Messner becomes the first climber to climb three eight-thousanders in one season. In spring, he climbs Kangchenjunga (8598 meters), followed by Gasherbrum II (8035 meters) and Broad Peak (8048 meters). A December attempt on Cho Oyu (8188 meters) with Hans Kammerlander is unsuccessful.

1983 In spring, Messner, Hans Kammerlander, and Michl Dacher make another attempt to climb Cho Oyu. This time they are successful.

Messner acquires Juval Castle, a thirteenth-century fortress built on a cliff between the Vinschgau and the Schnalstal in South Tyrol.

1984 In spring, Messner fails on the northeast ridge of Dhaulagiri. With Hans Kammerlander, he goes on to accomplish the Gasherbrum Traverse (Gasherbrum I and II in a single push), the first traverse of two 8000-meter mountains.

In summer, Messner begins restoration work on Juval Castle.

1985 Messner's father dies.

In April, Reinhold Messner and Hans Kammerlander climb Annapurna I (8091 meters) via the extremely difficult northwest face. Two weeks later they summit Dhaulagiri. In summer, Messner travels to Tibet, where he completes the pilgrims' walk around Mount Kailash. On the journey home, he learns of the death of his brother Siegfried, who was struck by lightning.

In the autumn, Messner moves into Juval Castle.

1986 In summer, Messner treks across eastern Tibet, from Kham to Lhasa, accompanied for part of the way by Sabine Stehle. He follows this trip with a fourth attempt on Makalu, with Friedl Mutschlechner and Hans Kammerlander. This time he is successful. Just three weeks later, he and Kammerlander stand on the summit of Lhotse, and Messner becomes the first person to climb all fourteen eight-thousanders. In December, Messner travels to Antarctica, where he climbs Mount Vinson (4897 meters), thus completing the "Seven Summits," the highest peaks on each of the seven continents.

Back in South Tyrol, he starts work on his small farm at Juval, employing organic methods of cultivation.

1987 Messner undertakes extensive treks in Bhutan and the Pamirs.

1988 Messner embarks on a solo expedition to Tibet in search of the yeti. Messner's partner, Sabine Stehle, gives birth to their daughter Magdalena.

1989 In spring, Messner leads an international expedition to climb the south face of Lhotse. The expedition is unsuccessful. He also undertakes further long treks in the Judaean Desert and in Patagonia. Messner and Arved Fuchs set off on their attempt to cross the Antarctic on foot.

1990 Arved Fuchs and Reinhold Messner cross the Antarctic via the South Pole, a distance of 2800 kilometers.

Reinhold Messner and Sabine Stehle's son, Gesar Simon, is born.

1991 Messner crosses Bhutan from east to west.

1992 Messner travels to Ecuador, where he climbs Chimborazo (6310 meters). Later in the year he makes a south-to-north crossing of the Takla Makan Desert in China.

1993 Reinhold Messner and his brother Hubert make a diagonal crossing of Greenland on foot, a distance of 2200 kilometers.

1995 Reinhold and Hubert Messner fail in their attempt to cross the Arctic Ocean from Siberia to Canada. In summer, Messner falls off a wall at Juval Castle and suffers a serious injury to his foot. Juval is opened to the public as a museum.

Messner's mother dies.

1998 Messner travels to the Altai Mountains in Mongolia and the Puna de Atacama in the Andes.

1999 Messner travels to the Thar Desert in India.

In June, he wins a seat in the European Parliament and joins the Greens/European Free Alliance (Greens/EFA) political group. He is a member of three standing committees: Regional Development, Transport and Tourism, and Agriculture and Rural Development. He is also a member of the delegation for relations with the countries of South Asia, and the South Asian Association for Regional Cooperation.

2000 In July, Messner leads a small expedition to Nanga Parbat, crossing the Diamir Glacier and climbing the north face of the mountain to the north summit. However, dangerous snow conditions on the summit ridge force him and his partner to abandon their attempt.

2001 Reinhold Messner and Sabine Stehle's daughter Anna Judith is born.

2002 In the UN International Year of Mountains, Messner visits indigenous mountain people in the Andes and climbs Cotopaxi (5897 meters) in Ecuador.

In June, the Messner Mountain Museum (MMM) Dolomites opens on Monte Rite near Cortina.

2003 Messner travels to Everest and Nanga Parbat to mark the fiftieth anniversary of the first ascents of the mountains. He attends the inauguration of the Günther Messner School in the Diamir Valley.

Messner develops a concept for MMM Firmian in Bozen, which will become the centerpiece of the Messner Mountain Museum.

2004 In spring, Messner makes a solo crossing on foot of the full length of the Gobi Desert in Mongolia.

On the occasion of the two hundredth anniversary of the first ascent of the Ortler, MMM Ortles is opened in Sulden.

2005 Messner visits the Tuwa nomads in Mongolia. After trekking around Nanga Parbat, he learns that a group of local mountain guides have found the remains of a climber on the Diamir Glacier. Messner identifies a boot as belonging to his brother, Günther, who was killed on the mountain in 1970. The boot and foot bones are taken back home to South Tyrol. The rest of Günther Messner's remains are cremated at the foot of the Diamir Face.

Messner sets up the Messner Mountain Foundation (MMF), which aims to support mountain people worldwide.

2006 The Messner family travels to Antarctica, in the footsteps of Shackleton.

Reinhold Messner opens MMM Firmian, the centerpiece of his museum project.

The Messner family travels to Nanga Parbat, where they view the work and the progress made by the Messner Mountain Foundation projects.

In November/December Messner leads a group across the Hielo Continental Norte (Northern Patagonian Ice Field).

2007 Messner and his son, Simon, climb in the Tassili Mountains in the Sahara.

2009 Messner collaborates with Joseph Vilsmaier on the film *Nanga Parbat*.

Messner climbs in Wadi Rum, Jordan, and travels to Namibia.

Sabine Stehle and Reinhold Messner marry.

2010 Messner travels to the Amazon. He then undertakes a trek to the holy mountain of Machapuchare in Nepal and joins the Himalaya Film Expedition.

2011 Messner climbs in the mountains of Sinai and travels to Easter Island. MMM Ripa, the fifth of Messner's museums, opens in Bruneck.

2012 Messner travels to Costa Rica. He also climbs the Gran Pilaster on the Pala di San Martino (2987 m) in the Dolomites.
Andreas Nickel's documentary film *Messner* is released.

2013 Film work for *Messner's Himalayas* in Nepal and Pakistan.
First ascent of Diagonale on the Geisler.

2014 Messner goes on a filming expedition to Ladakh.
The sixth and final Messner Mountain Museum, MMM Corones, opens. The museum is located at 2275 meters on the Kronplatz, above Bruneck, and is dedicated to the most difficult and demanding big wall climbing in the world.

TRANSLATOR'S NOTE

I was sixteen years old and had been climbing for just over a year when Reinhold Messner's first book, *The Seventh Grade*, was published in English in 1974. The book consists of a series of essays in which Messner describes his alpine climbs, interspersed with passages in which he formulates his philosophy on climbing and explains how he trained and prepared for his climbs. Messner advocated a bold new approach that limited the use of artificial aids and adopted the "by fair means" motto, a phrase first coined by the Victorian mountaineer A. F. Mummery. The book was subtitled *Most Extreme Climbing*, and Messner's passion for pushing the limits was obvious. For me, a self-confident teenager aspiring to test my own limits and climb my first "extreme," *The Seventh Grade* made compelling reading.

Two years later I packed my climbing gear in a rucksack, and some books and clothes in a borrowed suitcase and left home, bound for adulthood, a gritstone apprenticeship, and a degree course in German at Sheffield University. Climbing on the crags of the Peak District fed my obsession for short, hard routes done in the best possible style—no bolts, and on sight—while my degree course kindled a lifetime's interest in all things Germanic and led to my first translation assignments for *Mountain* magazine and Diadem Books.

In 1978, Messner and Peter Habeler rocked the climbing establishment by climbing Everest without the use of supplemental oxygen. I was on a teaching placement in Austria at the time and still remember the criticism they endured both before and after the climb, and the media frenzy their ascent provoked. Amongst the Austrian climbing scene, there was a feeling that something extraordinary had been achieved, although the true magnitude of the achievement was difficult to comprehend.

I first met Messner at the Alpine Club Symposium held at Sheffield Hallam University in 1999. In a pitch-perfect address, Messner reaffirmed his "passion for testing the limits" and reiterated his purist approach to adventure. He warned us against trying to make mountains safe, reminding us that "mountaineering is about risk" and that "a mountain without danger is not a mountain but something else." The following year I resigned my safe, pensionable full-time teaching post and moved to the Lake District to be closer to the mountains.

While working on the translation of this book, I was fortunate to be given the opportunity of visiting the Dolomites on a research trip. Splitting the week between museums and mountains, I hiked around the Sassolungo; took a trip to the "Messner Klettergarten" at the Zanser Alm in the Villnöss Valley, where Messner's father taught him to climb and which he describes as "the most beautiful place on Earth"; strolled along the Günther Messner Memorial Trail; and basked in the sunshine on the Kleine Fermeda.

The cultural highlight of the trip was a visit to MMM Firmian at Sigmundskron Castle near Bozen, the centerpiece of the Messner Mountain Museum project. At the entrance to the museum there is a plaque on the wall with a list of mountaineering greats. There are many familiar names on the list—Ricardo Cassin, Anderl Heckmair, Walter Bonatti, Joe Brown, Doug Scott, Royal Robbins, and Peter Habeler, to name just a few—and space for several more. There are also a few notable omissions. This is because one of the conditions (set by Messner) that mountaineers must fulfill in order to be included in this exclusive roll of honor is that they must have survived the experience, i.e., they must have reached the age of seventy at least.

Forty years have now passed since I read *The Seventh Grade*, and this year Reinhold Messner will celebrate his seventieth birthday. The name of the greatest mountaineer in history will be added to that list.

My job is now done, but before I pack a rucksack and go climbing. I would like to thank Barbara Scott and Tom Tarantino for their generosity, Ruth Ennemoser for her hospitality, Mary Metz for being such a nice person to work with, and my partner, Amanda, for her love and support and gentle criticism of my draft manuscript.

Tim Carruthers
Cumbria, May 2014

SPONSORS' NOTE

In hindsight maybe it was meant to be. We had planned a trip to the Dolomites, using a visit to each of the five (now six) Messner Mountain Museums (MMM) as an organizing device.

We knew that MMM Juval, the castle that is one of Reinhold Messner's residences, was closed during the summer but we had written six months beforehand, hoping that someone might be willing to make an exception for us.

Yes, we could visit. We showed up at 9:30 a.m., rang the bell, and, lo and behold, Reinhold Messner himself came down to the gate and let us in.

He was welcoming, open, and quite generous with his time. We figured that he must have been bored to be willing to see us, but we were going to take whatever we could get.

Mr. Messner told us how he had come to buy the place in the 1980s. He had hired an agent to represent him, fearing that if the seller knew that he wanted it, the price would only go up. Instead, his agent found that the seller was unwilling to sell to just another guy with money; he wanted to sell to a South Tyrolean, and, in fact, there was only one South Tyrolean he wanted to sell to and that was Reinhold Messner. It made for a happy transaction.

He showed us around. No, we didn't want to climb the rope ladder in the tower, but we did want to see the holds that people say he had installed in his living room so he could climb upside down across its ceiling.

He laughed, they weren't there . . . but they were in the basement and he showed them to us.

We are so grateful to the people who made our visit possible: Mrs. Uta Seeber, Petra Überbacher, and Mr. Erwin Domanegg.

And we thank you, Mr. Messner, and wish you a Happy 70th Birthday.

Barbara and Tom
September 17, 2014

ABOUT THE AUTHORS AND TRANSLATOR

Reinhold Messner, born in 1944 in Villnöss South Tyrol, is the most famous mountaineer and adventurer of our time. He has accomplished about one hundred first ascents, climbed all fourteen eight-thousanders, and crossed the Antarctic, Greenland, Tibet, and the Gobi and Takla Makan deserts on foot. After serving a term as a member of the European Parliament, he now devotes much of his time and energy to his Messner Mountain Museum (MMM) project and to his foundation, the Messner Mountain Foundation (MMF), which aims to support mountain people worldwide.

For more about the author, visit www.reinhold-messner.de.

Thomas Hüetlin was born in 1961 and grew up between Munich, Tegernsee, and Lech am Arlberg. He has worked for fifteen years as a reporter for *Spiegel* magazine.

Tim Carruthers, born in 1958, has been a climber since the age of fourteen. He has traveled and climbed extensively in Europe, the United States, and South America. He worked as a teacher of modern foreign languages until 1999, when he resigned his teaching post and moved to the mountains. Tim has translated more than a dozen books, including works by Hermann Buhl, Heinrich Harrer, Anderl Heckmair, and five other books by Reinhold Messner. He lives on a farm in the English Lake District.

MOUNTAINEERS BOOKS is a leading publisher of mountaineering literature and guides—including our flagship title, *Mountaineering: The Freedom of the Hills*—as well as adventure narratives, natural history, and general outdoor recreation. Through our two imprints, Skipstone and Braided River, we also publish titles on sustainability and conservation. We are committed to supporting the environmental and educational goals of our organization by providing expert information on human-powered adventure, sustainable practices at home and on the trail, and preservation of wilderness.

The Mountaineers, founded in 1906, is a 501(c)(3) nonprofit outdoor activity and conservation organization whose mission is "to explore, study, preserve, and enjoy the natural beauty of the outdoors." One of the largest such organizations in the United States, it sponsors classes and year-round outdoor activities throughout the Pacific Northwest, including climbing, hiking, backcountry skiing, snowshoeing, bicycling, camping, paddling, and more. The Mountaineers also supports its mission through its publishing division, Mountaineers Books, and promotes environmental education and citizen engagement. For more information, visit The Mountaineers Program Center, 7700 Sand Point Way NE, Seattle, WA 98115-3996; phone 206-521-6001; www.mountaineers.org; or email info@mountaineers.org.

Our publications are made possible through the generosity of donors and through sales of more than 500 titles on outdoor recreation, sustainable lifestyle, and conservation. To donate, purchase books, or learn more, visit us online:

**MOUNTAINEERS
BOOKS**

1001 SW Klickitat Way, Suite 201
Seattle, WA 98134
800-553-4453
mbooks@mountaineersbooks.org
www.mountaineersbooks.org

LEGENDS AND LORE SERIES

THE LEGENDS AND LORE SERIES honors the lives and adventures of mountaineers and is made possible in part through the generosity of donors. Mountaineers Books, a nonprofit publisher, further contributes to this investment through book sales from more than 600 titles on outdoor recreation, sustainable lifestyle, and conservation.

We would like to thank the following individuals for their charitable support of Legends and Lore:

- Anonymous
- Alex Bertulis
- Tina Bullitt
- Tom and Kathy Hornbein*
- Dianne Roberts and Jim Whittaker
- William Sumner
- Doug and Maggie Walker

*With special appreciation to Tom Hornbein, who donates to the series all royalties earned through the sale of his book, *Everest: The West Ridge*.

You can help us preserve and promote mountaineering literature by making a donation to the Legends and Lore series. For more information, benefits of sponsorship, or how you can support future work, please contact us at mbooks@mountaineersbooks.org or visit us online at www.mountaineers books.org.

MOUNTAINEERS BOOKS

The Legends and Lore series was created by Mountaineers Books in order to ensure that mountain literature will continue to be widely available to readers everywhere. From mountaineering classics to biographies of well-known climbers, and from renowned high-alpine adventures to lesser-known accomplishments, the series strives to bring mountaineering knowledge, history, and events to modern audiences in print and digital form.

Distinctive stories in the Legends and Lore series include:

Free Spirit
Reinhold Messner
"*Free Spirit* condenses a lifetime of climbing achievement into 288 pages of nerve-wracking reading."
—Mark Twight

Freedom Climbers
Bernadette McDonald
"A brilliantly crafted tale of mountain and political adventure that reveals a golden era in Himalayan climbing that was as glorious as it was tragic."
—Sir Chris Bonington

The Roskelley Collection: Nanda Devi, Last Days, and Stories Off the Wall
John Roskelley
"From his early days struggling with team dynamics to his grown-up climbs in the Himalayas, Roskelley doesn't hold back."
—*Climberism* magazine

Minus 148°: The First Winter Ascent of Mount McKinley
Art Davidson
"This finely crafted adventure tale runs on adrenaline but also something else: brutal honesty."
—*Wall Street Journal*

MOUNTAINEERS BOOKS